S0-CAS-426

Analytical Aspects of
Environmental Chemistry

6922352X
CHEM

CHEMICAL ANALYSIS

A SERIES OF MONOGRAPHS ON
ANALYTICAL CHEMISTRY AND ITS APPLICATIONS

Editors

P. J. ELVING   J. D. WINEFORDNER

Editor Emeritus: I. M. KOLTHOFF

Advisory Board

Fred W. Billmeyer, Jr.
Eli Grushka
Barry L. Karger
Viliam Krivan

Victor G. Mossotti
A. Lee Smith
Bernard Tremillon
T. S. West

VOLUME 64

CHEM
Sep/ae

A WILEY-INTERSCIENCE PUBLICATION

JOHN WILEY & SONS
New York • Chichester • Brisbane • Toronto • Singapore

SD3/14/83PD

Analytical Aspects of Environmental Chemistry

QD75
.3
AS
1983
CHEM

DAVID F. S. NATUSCH

Liquid Fuels Trust Board
Wellington, New Zealand

PHILIP K. HOPKE

Institute for Environmental Studies
University of Illinois, Urbana

A WILEY-INTERSCIENCE PUBLICATION

JOHN WILEY & SONS

New York • Chichester • Brisbane • Toronto • Singapore

Copyright © 1983 by John Wiley & Sons, Inc.

All rights reserved. Published simultaneously in Canada.

Reproduction or translation of any part of this work
beyond that permitted by Section 107 or 108 of the
1976 United States Copyright Act without the permission
of the copyright owner is unlawful. Requests for
permission or further information should be addressed to
the Permissions Department, John Wiley & Sons, Inc.

Libary of Congress Cataloging in Publication Data:

Main entry under title:

Analytical aspects of environmental chemistry.

(Chemical analysis; v. 64)
"A Wiley-Interscience publication."
Includes index.
Contents: Chemical speciation / Robert S. Braman—
Capillary gas chromatography in the analysis of environ-
ment / Milos Novotny—Recent advances in combined gas
chromatography / Philip W. Ryan—[etc.]
1. Chemistry, Analytic. 2. Environmental chemistry.
I. Natusch, David. II. Hopke, Philip K., 1944–

QD75.3.A5 1983   543   82-13518
ISBN 0-471-04324-9

Printed in the United States of America

10 9 8 7 6 5 4 3 2 1

CONTRIBUTORS

Robert S. Braman
Department of Chemistry
University of South Florida
Tampa, Florida

G. E. Cabaniss
Department of Chemistry
The University of North Carolina
Chapel Hill, North Carolina

S.-G. Chang
Lawrence Berkeley Laboratory
University of California
Berkeley, California

R. L. Dod
Lawrence Berkeley Laboratory
University of California
Berkeley, California

L. Gundel
Lawrence Berkeley Laboratory
University of California
Berkeley, California

D. T. Harvey
Department of Chemistry
The University of North Carolina
Chapel Hill, North Carolina

Philip K. Hopke
University of Illinois
Institute for Environmental
    Studies
Urbana, Illinois

R. W. Linton
Department of Chemistry
The University of
    North Carolina
Chapel Hill, North Carolina

T. Novakov
Lawrence Berkeley
    Laboratory
University of California
Berkeley, California

Milos Novotny
Department of Chemistry
Indiana University
Bloomington, Indiana

Philip W. Ryan
Systems, Science and
    Software
La Jolla, California

PREFACE

The objective of this volume when it was first being organized was to highlight analytical techniques that would become increasingly important in solving environmental chemistry problems primarily associated with increasing the specificity with which a quantitative analysis can be performed. In the period that has passed, additional developments have been completed and many of these methods have become more and more widely used.

Many early environmental studies focused on elemental concentrations. The developments reported here permit the identification and quantification of specific compounds. The importance of these improvements is that they permit us to obtain a much more detailed understanding of the complex chemistry that takes place in environmental systems. Since the effects of various elements are intimately related to their chemical speciation, it is the ability to determine the amounts of specific compounds that permits a better estimation of the effects of discharges of various elements such as lead, tin, mercury, and arsenic.

With the improvements in chromatographic techniques, very complicated mixtures, typical of the environment, can be separated and the components identified using mass spectrometry. The broad spectrum of natural and anthropogenic compounds possibly present in a sample makes this ability to separate and identify essential to the understanding of the system under study, whether it be photochemical smog, the chlorination of wastewater effluents, or the weathering of spilled oil.

Another of the important new aspects of environmental analysis has been the study of surfaces. The ability to closely examine the chemical nature of a surface or to scrutinize individual particles substantially increases our knowledge of the species present at the point of contact with the environment. Not only can we examine the elemental concentrations of single particles but we can examine the structure of the particle as we peel away the surface, identifying changes in composition with depth into the particle. The ability to identify different chemical states of an element in a particle and to compare those results with bulk analysis techniques indicates that what is commonly found by such bulk analyses is not really what is present in the particle or that the surface concentration is much higher than the apparent bulk concentration. All this detailed information gives us a much better understanding of the chemistry of the environment.

Finally, environmental chemists have realized that it is often difficult to even identify all of the quantities that need to be measured in order to characterize an environmental system. Yet we have methods that produce sets of data so large that it is difficult for an individual to directly extract information from them. The application of various statistical techniques then becomes a valuable tool to interpret the processes that are occurring. These techniques also can be a valuable part of a quality assurance program to identify outliers and examine the data for systematic biases. It must be realized, however, that the techniques require someone knowledgeable about the physical and chemical processes of the real world to ensure that these interpretations are realistic.

We wish to thank the authors who have contributed to this volume for their patience and diligence in seeing it through to completion. Thanks are particularly due to Barbara Cummins for assisting in the preparation of several of these manuscripts.

DAVID F. S. NATUSCH

Wellington, New Zealand

PHILIP K. HOPKE

Urbana, Illinois
November 1982

CONTENTS

ix

Analytical Aspects of
Environmental Chemistry

CHEMICAL SPECIATION

ROBERT S. BRAMAN

Department of Chemistry
University of South Florida
Tampa, Florida 33620

1. CHEMICAL SPECIATION

1.1. Introduction

The term "speciation" has come into a certain degree of use in the past few years in connection with environmental analysis. As applied, the term means identification of inorganic, organometallic, or organic compounds actually present in the environment. This is not always an easy (or even a possible) task, as will be developed later. The term speciation really draws attention to the distinction between measuring the total concentration of an element and measuring the concentration of each of its chemical forms.

Chemists have, of course, been measuring the concentration of specific chemical compounds or ions for many years, if not from the very start of chemical analysis. Some of the first and most widely known examples of chemical analyses on environmental samples provide a "speciation." Particularly there comes to mind the water analyses for NO_3^-, NO_2^-, NH_4^+, and "amine type nitrogen" in which nitrogen is reasonably well divided into its most important environmental forms, if not all the specific molecular forms present. In air we have the CO, CO_2, and "total hydrocarbon" type analyses, which are also speciation to some extent for carbon. Gas chromatographic analyses for specific organic compounds is speciation of a type, and many thousands of applications have been made.

This chapter is not intended to be an exhaustive discussion of environmental analysis methods that have a speciation capability realized or implied. What is intended is to discuss the problem of determining low concentrations of specific metal complexes or organometallic forms of elements of current interest. Included will be some general comments on this type of analytical problem and air and water analyses as specific problems followed by a review of some specific elements of interest.

1.2. Importance of Speciation Studies

With the advent of x-ray fluorescence spectroscopy, improved emission spectroscopy, neutron activation analysis, much improved multielement emission spectroscopy, and atomic absorption spectroscopy, the collection of total element data has been so much facilitated that these methods have been widely used in environmental analyses. As a result, total element data dominate the literature, at least for the metallic and semimetallic elements. Maximum permissible concentrations of various elements are given in terms of total element concentrations. Most of the data on total element composition has been used with reasonably good effects in environmental studies or in monitoring. It is certain that total element concentration data are sufficient for some elements—perhaps, eventually, for all except a very few when pollution monitoring is involved. Nevertheless, this has drawn attention away from studying the activities of specific forms of elements in the environment, which is an unfortunate state of affairs that has only recently been corrected.

Speciation studies are important for several reasons. The biological activity, for example, toxicity, of an element can vary widely from compound to compound, and it is important to know if and how toxic forms are produced or handled in biological systems. Because of differences in the important physical properties of volatility and solubility for various chemical forms of an element, one or several may dominate physical exchanges in global geochemical cycles. Similarly, chemical properties may strongly affect rates of decomposition and rates of local transportation of various elements (important in cases of pollution) as well as rates of transportation on a global scale.

A matter that has been given less attention here than it deserves is the need for monitoring on a global basis in nonpolluted locations the specific forms of elements—especially those having reasonably toxic and environmentally mobile forms. Almost no data of this type exist on most elements. Elements suggested as most important to study are sulfur (well known to be mobile), arsenic, mercury, and selenium, but other likely biologically active elements such as tin, lead, tellurium, and antimony should be added to the list. What must be determined is whether the several forms of these elements remain in more or less constant ratio to each other, indicating an approximate equilibrium condition in global cycles, or if any form is increasing versus the others on a local or on a global basis, a situation that could have rather grave consequences.

A related and similar speciation problem is the study of specific "point type" pollution sources for speciation type of information. Point sources here could be physically large-sized areas such as traffic systems, sanitary landfill areas, estuaries, and embayments.

Despite the transport of pollution from sources, by the wind in the case of air pollution or by river flow in the case of water pollution, there may be sufficient time for pollutants to interact with the biological and physical systems present and possibly come to a degree of equilibrium, a situation that usually means the buildup of a localized concentration of one or several forms of an element. Other examples of localized conditions are air entrapped in meteorological inversions and air partially trapped under woodland canopies. Isolated lakes are good examples where aqueous–sediment interface equilibria can be established. The anaerobic waters of certain marine areas such as the Curacau trench and embayments exhibiting poor tidal flushing are good examples of isolated saline waters. Although these localized areas can hardly be considered in an equilibrium state of the kind considered in global cycle modeling, nevertheless, they could approach equilibrium if chemical exchanges and reactions proceed at rates greater than losses to the exterior of the affected area. It would be instructive to determine whether a material balance of transported forms of an element can be made based on appropriate analyses.

The study of these isolated systems should prove interesting when compared to nonpolluted or nonisolated locations. A different speciation ratio of a particular element compared to that of the same element in ambient locations will probably be found depending on the prevailing pE and pH. In aqueous systems the pE (oxidation potential is really simply a factor related to the dissolved oxygen concentration) as well as the pH may be substantially different from that of nonisolated locations, for example, open ocean or ecologically "well-balanced" (nonentrophied) freshwater lakes.

Obviously, chemical forms of many elements are susceptible to oxidation and reduction reactions in the limits of conditions available in the environment. In the case of air, the range of pE values is limited because oxygen is omnipresent. A range of surface pH values is available. The chemical nature of air samples depends on the input of a certain compound or group of compounds from pollution sources if present.

Speciation studies and compositional type models such as pE–pH diagrams compliment each other. Results of environmental studies aid in better describing the real aqueous system present. This leads to a better interpretation of the pE–pH diagrams, which are, in fact, nothing more than intelligent guesses of what should be present.

Speciation studies are needed in several specific circumstances. Particularly important is their use in preparing environmental impact statements. When chemical pollution is involved, total element values may not be sufficient to indicate potential toxicity problems. In addition, when extrapolating from existing data and applying the results to a proposed new location being studied, the chemical, physical, and biological makeup of the new areas (i.e., pE, pH, water composition, temperature, hydrology, and biology) must be considered. If the overall chemical environment of the new area is only somewhat different from that of the prior study, the speciation could be radically different. Where environmentally mobile elements are important, speciation studies should be carried out prior to the completion of an environmental impact statement.

Speciation studies can influence the selection of environmental monitoring equipment. In some cases it may be more useful to test for a single compound than to obtain a total element value. This could be true, for example, if total element analyses are different because conversion of a particulate solid phase form into an analyzable form is difficult and requires complicating steps in the analysis procedure.

1.3. Strategy in Speciation Studies

Even in the instance of some elements for which reasonably good analytical methods exist, the determination of only total element present in a sample can be time-consuming and expensive. The prospect of analyzing for every determinable form of every element is simply not feasible.

The best compromise between knowing nothing about the chemistry of a particular element in the environment and trying to do the impractical is to do studies on a case-by-case basis and monitor by using the easier total element analyses where available. Thus it should suffice to analyze once for the several forms of arsenic in a lake being polluted (perhaps by an organic arsenical), in the air above the lake, and in the adjacent area, sediments, soil, plants, and other biota. After this, total element analysis of lake water and perhaps also of the air should suffice for monitoring.

Speciation studies may be partially repeated over long intervals after the first round to determine if any significant shifts in the ratio of chemical forms are occurring. In this regard, seasonal variations in speciation may be observed as a result of changes in temperature, degree of sunlight, ground cover, and biological activity.

The speciation of an element will probably be the same for locations

that have approximately the same types of pollution and makeup of the
ambient area. This can serve to guide the selection of appropriate analytical methods. For example, the range of sulfur compounds found near
paper mills should be the same no matter where the mills are located,
and a similar environmental pollution profile should be found.

1.4. Nature of Speciation as an Analytical Problem

The analytical problems encountered in environmental studies are, as
in other areas of analytical chemistry, tied to the objective of the intended
study. Use of the data obtained governs the analyses needed. A quick
survey may require an analytical method having a 10 ppb limit of detection and only for some empirically selected fraction of the element
present—such as "soluble element" defined as in water samples that pass
a 0.45-μm filter. A carefully done biological impact assessment could
require a 0.5-ppb limit of detection of the same element with some indication of its organometallic forms present.

Unless analytical work has been done on a similar problem, the environmental researcher may find the progress of his or her work limited
by the need for analytical methods development. This is not a very pleas-
ing situation especially for those who are much given to making and following research plans in an orderly fashion. Particularly upsetting is the
need often found to spend large amounts of time developing analysis techniques and proving their reliability and accuracy in field operations or in
analyses of real samples.

Generally, the analytical speciation problem has the following characteristics. (1) Samples are a complex mixture of compounds often including elemental forms. (2) Part or all of the mixture can be sensitive to
air oxidation or chemical reduction during ordinary sampling and sample
processing. (3) Part or all of the mixture can be sensitive to pH changes.
(4) The range of volatility of the chemical forms of an element can be
from that of a solid to that of a gas. (5) Polymers and complexes may be
present. (6) Total analyte is likely to be less than 1 μg per sample taken
and often less than 1 ng for a convenient sample size. (7) Sometimes the
presence of the components sought has not been proven in real environmental samples despite a lot of intelligent speculation or opinion and even
good laboratory model chemical systems work. (It is particularly at this
point that the analytical chemist working on environmental chemistry

finds himself crossing fields of scientific discipline.) (8) Finally, there may be few or no good ways to verify the analytical method since the one just developed is the only one available.

The usual logical process in developing an analytical method involves first a statement of the analysis needed followed by application of the properly selected analytical procedure. Unfortunately, in speciation work the statement of the problem is the major part of the problem. One often finds oneself restating the problem during its study. This is a common phenomenon in science but certainly one that again underlies the fact being emphasized here: Analyses for speciation studies are not routine and should not be treated as such by anyone, analytical chemists or scientists of other persuasions.

The balance of this chapter on speciation will concern itself with first general approaches to speciation in air and in water. This will be followed by a review of speciation of selected elements that have caught the interest of researchers in recent times.

2. SPECIATION IN AIR ANALYSIS

2.1. Optical Absorption Methods

No studies on the environmental distribution of the several forms of a particular element have been reported using optical absorption methods in either aqueous or gas samples. The very low concentrations of compounds encountered in the environment simply cannot be detected by the several types of instruments now available or under study. Nevertheless, considerable progress is being made and a review of current capabilities is in order.

Optical absorption methods do have the potential advantage of good selectivity for individual molecules particularly in the IR region of the spectrum. They also possess the obvious advantage of not requiring a chemical modification or physical transfer (e.g., a separation or chemical reaction) prior to analysis. Of all the air analysis sample acquisition techniques, optical absorption methods have the highest collection efficiency (100%) and perturb the nature of the sample the least.

A very substantial amount of interest has gone into developing optical absorption methods of analysis for trace materials in the environment.

Applications have included stack monitoring, long path infrared monitoring using ordinary sources, and analysis and monitoring using laser sources. For more background, the reader is referred to the book on air pollution analytical methods edited by Stevens and Herget (1) and a book on laser monitoring by Hinkely (2).

Early laser devices were of the fixed wavelength type without much capability of wavelength selectivity. The availability of tunable lasers has greatly enhanced both selectivity and ability to perform multicomponent analysis. Hinkley (3) has compared the capabilities of the several type of tunable lasers. Typical applications have been to the detection of specific important pollutants. Table 1A, with data from Hinkley and Calawa (4) illustrates current detection limits attainable with a diode laser instrument albeit over a 2-k path length. Table 1B data from Pitts and co-workers (5, 6) give detection limits for a He–Ne laser source, Michelson interferometer–computer type instrument. The interferometer approach, which is in part a Fourier transform infrared spectrometer, has the ad-

TABLE 1.　Detection Limits for Laser Source Optical Absorption Analysis Systems

Compound	λ (μm)	Sensitivity (ppb)[a]
A.　Diode Laser (4)		
CO	4.74	0.3
O_3	4.75	3.0
NO	5.31	1.5
SO_2	8.88	15
C_2H_4	10.54	0.8
B.　He–Ne Laser, FT–IR (5, 6)		
NH_3	931 (μm^{-1})	4[b]
HCHO	2779	6
HCOOH	1110.3	2
HNO_3	896	6
HNO_2	853	4
O_3	1055	10
PAN	1162	3
H_2O_2	1250	8
CH_3OH	1033	2

[a] Path length is 2 km.
[b] Path length is 0.9 km.

vantage of a short wavelength scanning time. A computer is needed to convert the interferogram to its Fourier transform, a conventional IR spectrum. The need for a computer is not a disadvantage; the computer will be needed for resolution of data obtained from mixtures of compounds also. Green and Steinfield (7) have studied the use of a CO_2 laser for monitoring mixtures of organic vapors in air. Minimum detectable concentrations were in the parts-per-million range for a 10-m pathlength. Some typical values are: benzene, 25 ppm; *t*-butanol, 14 ppm; vinyl chloride, 6.4 ppm; *d*-chloridifluoromethane, 1 ppm; and ethyl acetate, 4.2 ppm. These are concentrations at which a 20% error in measured absorption is observed for the single compound alone. When a mixture of compounds is to be analyzed, a multiple wavelength measurement technique is used. A computer programmed least-squares analysis of the data is used to resolve the mixture.

In summary, it appears that long path optical absorption methods do not yet demonstrate sufficient sensitivity for any analysis purposes at ambient concentrations below parts per billion. Nevertheless, good applications have been made to analyses of individual pollutants with detection limits approaching this. Through the use of multiple wavelength measurements and computer programmed solution of simultaneous equations, the capability of multiple component analysis exists. It is a virtual certainty that if detection limits are reduced to below 1 ppb by state-of-the-art advances, more speciation applications will become possible but at the cost of a greatly increased complexity of detected sample components.

2.2. Preconcentration, Separation, and Detection

2.2.1. *Cryogenic and Cold Trapping*

Air components in vapor form often are present in concentrations well below those that permit direct analysis. Cold trapping and cryogenic trapping (which is simply cold trapping with liquid air, oxygen, or nitrogen) can be used to collect these volatile compounds. The usual approach is to cool a tube packed with some suitable solid material and draw air through the tube by means of a pump. A number of problems occur with this type of sampling, the chief of which are water removal and accuracy of determining sample air volume. Ten liters of air at 25°C will contain 0.25 mL of water at 100% relative humidity. If more than a few liters of air are to be sampled, a water-removing pretrap or a large drying tube is needed to avoid plugging the cold sampling trap with ice. The trapping of water with the sample can obstruct the air flow through the sampling tube, an effect that changes with sampling time. This makes the accurate determination of the sampled air volume difficult. The use of an in-line wet test meter is suggested.

Another difficulty with cryogenic trapping is the rapid absorption of major air components, CO_2 and O_2, onto the solid surface of the absorbent used. High surface area solid sorbents such as Chromosorb 102 absorb the most. Two effects result from this: The absorption of air onto the solid surface can be so rapid at first as to produce a greater than pumped flow rate into the cooled absorption device. Upon warming, the liquidified gases produce a very high flow rate out of the sampler, sometimes physically expelling the frozen analyte as an aerosol or particulate spray. These problems probably limit the wide application of cold trapping in field analysis or monitoring especially when 10–20 L of air or more must be processed. Altshuller (8) has reviewed cryogenic trapping methods.

Cryogenic trapping has often been used to concentrate comparatively small sample volumes, 50–500 mL, prior to analysis by gas chromatography (9–11). Cryogenic trapping is required for preconcentration of the light hydrocarbons. Even so, there can be problems. This writer has found that methane (b.p. $-164°C$) was not even quantitatively absorbed out of a He carrier gas stream at 100 mL/min on glass beads at liquid nitrogen temperatures. A high surface area sorbent (Chromosorb 102) was needed to obtain quantitative trapping of methane.

Physical absorption onto solid sorbents of a variety of types has been used for preconcentration of vapors out of air prior to analysis. Physical absorption at ambient temperatures avoids water absorption problems and can provide good preconcentration. Analysis procedures usually consist of sample absorption, rapid heating of the sorbent cartridge in a GC carrier gas feed stream followed by separation and analysis in the usual fashion. Rapid and quantitative desorption is the key to good results. It is probably best to desorb from a high surface area trap into a small surface cryogenic trap followed by separation and analysis. This technique reduces the size of the sample volume and thus improves separation.

A reasonably large number of solid sorbents have been studied for preconcentration of vapors out of air at ambient temperatures. In every case where physical absorption is the cause of preconcentration, chromatographic behavior of the absorbed material is observed. Compounds slowly migrate down an absorption tube during use and exhibit a retention volume or, perhaps more exactly, a breakthrough volume. Breakthrough should be a linear function of sorbent weight but an exponential function of temperature. Since each different absorbing material under consideration has its own surface character, each should be evaluated for breakthrough volume at several temperatures near the desired ambient sampling temperature using the compounds to be preconcentrated. From these data, one can determine the maximum sampling volume. Some breakthrough will be observed prior to the saturation of the absorbed component. The maximum useful retention volume should probably be taken when the detected compound is present at 5% of its maximum or feed value. Pellizzari et al. (12) have reported a technique for evaluating sorbents for a number of organic compounds. Parsons and Mitzner (13) have reported on the use of Tenax for sample preconcentration. Dravnieks et al. (14) have evaluated Chromosorb 102 for absorption of organic vapors. Tenax-GC is specified by both the U.S. Environmental Protection Agency (15) and the National Institute of Occupational Safety and Health (NIOSH) (16) for collection of vapor constituents from air as well as volatile organic compounds from water (15). NIOSH also recommends activated charcoal for organic solvents and vinyl chloride determinations (16). The use of charcoal adsorption techniques and their limitations has been reviewed by the National Academy of Sciences (17). Both Tenax-GC and porous polyurethane foam have been found to be efficient adsorbents for collection of atmospheric organochlorine compounds (18). There are considerable problems associated with the physical adsorption of low boiling point compounds. An effective alternative to physical adsorption is preconcentration by chemical reaction or reversible chemisorption, which are treated in the next section.

2.2.2. *Preconcentration by Reversible Chemisorption or Chemical Reaction*

As pointed out in the preceding section, preconcentration by purely physical absorption onto high surface materials is not suitable for the large volume preconcentration of the more volatile trace compounds. In many important speciation studies, the components sought are quite volatile at ambient temperatures. For example, $(CH_3)_3As$ (b.p. 70°C), $(CH_3)_4Sn$

(b.p. 78°C), CH_3SH (b.p. 6°C), $(CH_3)_4Pb$ (b.p. 110°C), and $(CH_3)_2Hg$ (b.p. 96°C), all important biomethylated forms, are present in the environment in very low concentrations, below 10 ng/m^3 in most cases and, despite the excellent detection limits of analysis methods suitable for their detection, preconcentration of 100–1000 L or more is needed. Tenax-GC and other sorbants are not suitable for this degree of preconcentration. The only alternative to cryogenic trapping is preconcentration by chemical reaction or by a reversible chemisorption.

The application of preconcentration methods in analyzing environmental samples depends on a number of factors: concentration of the analyte, chemical nature of the analyte, vapor pressure of the analyte at the sample ambient temperature, and the detection limit of the detector used. Most of these influence the size of the sample that must be taken to detect the compound sought.

Chemical reaction approaches are among the oldest approaches involving simultaneous preconcentration and analysis for trace compounds in air. Usually a chemical that produces a color reaction with the analyte is absorbed onto a solid surface in a small absorption tube. This type of sampler can be made very specific by careful selection of the color reagent and chemical condition of the support material. Sensitivity is not great since colorimetric methods are used. The approach is satisfactory for detection of gross accidental contamination by certain common industrial compounds. Small hand-pumped devices are available from common chemical apparatus supply firms. For example, one firm had available detectors for H_2S (0–650 ppm), SO_2 (0–2700 ppm), CO_2 (0–40°C), NO_2 (0–55 ppm), Cl_2 (0–20 ppm), and CO (0–0.2%).

Although hardly in the same sensitivity range needed for speciation studies, the approach remains a good one. For extension to use at very low concentrations, the only change needed is an improvement in the detection limits of the analyzing device. Chemical reaction devices do not display the retention volume phenomenon of physical absorption-based devices. They should be capable of high volume preconcentration as long as the reaction capacity of the analyte specific chemical trapping reagent is not exceeded.

The sampling of air by chemically treated tapes or filters for H_2S represent the closest to speciationlike applications for specific reactions. Ozone is detected by observing the chemiluminescence on a silica gel surface treated with Rhodamine-B (19–21). Lead (22), silver (23–24), and mercury (24) treated filters have been used to detect low concentrations of H$_2$S in air.

Reversible chemisorption preconcentrators are devices in which much stronger absorption mechanisms are functioning to trap analytes than is the case in physical absorption. The term "reversible" implies that the analyte chemisorbed is capable of being desorbed without losing its identity, a factor necessary to speciation. Desorption can be effected by displacement with a more strongly chemisorbed compound, by thermal decomposition to a derivative compound that still retains the identity of the original sorbed analyte, or by chemical reaction to produce a compound that is not chemisorbed by the surface and that retains the identity of the sorbed analyte.

Apparently only a few examples have appeared of the reversible chemisorption type of preconcentration. Metal surfaces with specific activity for Hg compounds (26), methyltin compounds (27), methylarsines (28, 29), and sulfur compounds (30) have been reported. These are described in more detail in the sections on the specific elements involved. Nevertheless, as an example, trimethylarsine, dimethylarsine, methylarsine, and arsine vapors are absorbed onto clean silver-coated glass beads. The compounds cannot be desorbed directly by heating, but a warm solution of a low concentration of NaOH in water removes the arsines, apparently in the oxidized forms $(CH_3)_3AsO$, $(CH_3)_2AsO(OH)$, $CH_3AsO(OH)_2$, and $HAsO_2$. These may be reduced to the original arsines in the NaBH$_4$ reduction atomic emission spectroscopic detection method (31) and thus reactive desorption does not destroy speciation.

Recently a new approach has been developed to collect analytes on specific chemisorbant interior coatings of hollow tubes as preconcentrators (32). This technique preferentially removes volatile components from gas samples without the problems associated with filter reactions. The experimental results agree well with the mathematically modeled dynamics of the tubes. The diffusion coefficients of the chemisorbed gases can be determined with reasonable precision. The diffusion coefficients can be used to predict the molecular form or detect the presence of weak associations of detected molecules. It is a technique that can be applied to ambient air at commonly encountered concentrations and can be used for any analyte for which a rapid surface chemisorption reaction and reversible desorption or removal method is available. The general approach is to preconcentrate gaseous analytes on a specific absorbing surface and particulate analytes on a filter or packed tube. This is followed by an analysis of either or both preconcentrators. To date, only a few specific applications have been made (32, 33), but the potential applicability is great.

3. SPECIATION IN WATER ANALYSIS

3.1. Ion Selective Electrodes and Anodic Stripping Voltammetry

Most electroanalytical methods are used to determine total elemental composition of samples. Anodic stripping voltammetry (ASV) using the differential pulse stripping mode has excellent detection limits for a number of metals as described by Copeland and Skagerboe (34). For Cd, Cu, and Pb detection limits are in the $10^{-10}-10^{-12}$ g/L range. The only speciation of sorts obtained has been a designation of "labile" versus "strongly bound" metals in water samples.

Chau and Lum-Shue-Chen (35) designated labile or free metal ion results obtained by analyses of samples without any special treatment. Results from samples analyzed after nonoxidative acid digestion were designated total metal ion. The difference was designated strongly bound metal ion. Presumably the acid treatment helps to release metal ions from organic macromolecules.

One can only obtain limited speciation information by the direct use of ASV. A considerable improvement will be application of ASV to the analysis of samples after separation on a high performance liquid chromatograph or separation by chromatography. Sensitivity of ASV is sufficient for this type of use and the technique also has the considerable advantage of multielement detection capability. The major problem will be speed of the ASV experiment. Perhaps one of the best applications of ASV is studying model environmental systems as has been done by Petrie and Baier (36) for inorganic lead.

Ion selective electrodes are widely used for batch analyses or for monitoring. Each electrode is reasonably specific for detection of a single ion or it can be made so by appropriate treatment of the sample solution. This approach has some sense of speciation in that a specific chemical form is detected. By analyzing samples before and after treatment, oxidation for example, it is possible to differentiate between total and bound metal ion.

Frant (37) has reviewed electrodes applied to pollution analyses, Cd^{2+}, Pb^{2+}, CN^-, Cl^-, F^-, NO_3^-, K^+, Na^+, and other ions are detected in the 10^{-5}–10^{-7} range, approximately in the 10–100 ppb range. Buck (38) has reviewed the subject. The section on copper in this chapter gives information on a Cu^{2+} ion electrode that has been used down to 10^{-8}–10^{-9} M.

No electrodes appear to have been developed for organometallic cation detection, for example, $(CH_3)_2Sn^{2+}$. There appears to be no theoretical barrier to developing such an electrode but the detection limit would have to be very low indeed (10^{-12} M) to be used in natural water systems unless pollution were evident.

Ion selective electrodes have been found useful in determining the average formation constant of naturally occurring ligands as discussed in the section on natural complexes.

3.2. Preconcentration, Separation, and Detection

Extraction methods are widely used for the preconcentration of water samples prior to analysis by atomic absorption spectroscopy. Concentration limits of detection are improved by a factor of approximately 100, which is calculated by assuring that a 1-L sample is extracted into a volume of 50 mL and that a response improvement of four times is obtained because the analyte is in an organic solvent. Thus a metal that may be readily analyzed with good precision at 1 ppm in water directly can be analyzed with the same precision at 10 ppb, which is quite suitable for many purposes.

For speciation purposes, extraction must be followed by separation prior to analysis. It would appear that different ionic forms of the same element could be extracted and then separated but little work of this type has been reported. Most extraction procedures are designed for total element analyses. Specific procedures and applications are abundant. The most notable example of extraction with speciation is the work of Daugherty, Litchett, and Mushak (39) and Mushak, Dessauer, and Walls (40).

Arsenic(III) and organoarsenic compounds are converted under appropriate conditions to the diethyldithiocarbamate derivatives, extracted

into organic solvent and analyzed by gas chromatography. Obviously
other applications of this approach can be made, for example, in the
analysis of the tin compounds.

The preparation of volatile hydrides of certain trace metals in aqueous
solution by the use of $NaBH_4$ has been successfully used in a method for
their preconcentration prior to separation and determination of different
chemical forms. Applications of the $NaBH_4$ approach are summarized in
Table 2. Note that methyl compounds of many metals may be converted
to corresponding hydrides without loss of methyl groups or dispropor-
tionation. This approach is particularly effective in speciation work because the hydrides prepared retain the original speciation, are generally
volatile, quite insoluble in water, and thus can be efficiently removed
from the reaction chamber by gas scrubbing prior to collection in a cold
trap. A typical apparatus for this technique is shown in Figure 1. The cold
trap may be packed with glass beads or gas chromatographic column
packing material, the choice of which should be examined so as to avoid
chemisorption losses of trapped hydrides. Traces of metals in column

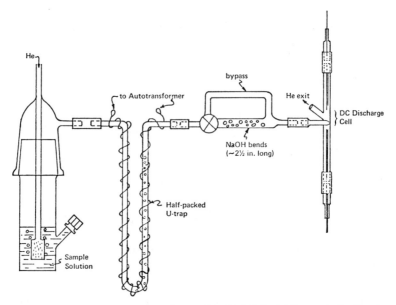

Figure 1.   Reaction chamber and U-trap for covalent $NaBH_4$ hydriding methods. Reprinted
with permission from R. S. Braman, D. L. Johnson, C. C. Foreback, J. M. Ammons, and
J. L. Bricker, *Anal. Chem.* **49,** 621 (1977). Copyright 1977 American Chemical Society.

TABLE 2. Applications of the NaBH₄ Reduction to Environmental Analyses

Element/Compound	Hydride Produced	Detector	Detection Limit	References
As(III,V)	AsH$_3$	AES	0.2 ng	31, 45
		AES	0.02 ng	49
		AA	3 ng	41
		AA	0.05 ng	46
CH$_3$AsO(OH)$_2$	CH$_3$AsH$_2$	AES	0.2 ng	31, 45
		ECD	0.4 ng	46
(CH$_3$)$_2$AsO(OH)	(CH$_3$)$_2$AsH	AES	1.0 ng	31, 45
		ECD	0.2 ng	46
(CH$_3$)$_3$AsO	(CH$_3$)$_3$As	AES	1.0	31
		ECD	15 ng	46
		FID	1 ng	46
Bi(III)	BiH$_3$	AA	8 ng	44
Ge	GeH$_4$	AA	270 ng	44, 46
		AES	0.18 ng	48
CH$_3$GeCl$_3$	CH$_3$GeH$_3$	AES	0.31 ng	47, 48
(CH$_3$)$_2$GeCl$_2$	(CH$_3$)$_2$GeH$_2$	AES	0.69 ng	47, 48
Sb(III)	SbH$_3$	AA	2 ng	41
		AES	0.35 ng	48
Se(IV)	H$_2$Se	AA	2 ng	41
		AES	5 ng	42
		AES	0.04 ng	50
Sn(IV)	SnH$_4$	AA	7 ng	44
		AES	0.1 ng	43
		FPD	1.3 pg	43
CH$_3$Sn^{3+}	CH$_3$SnH$_3$	AES	0.074 ng	43
		FPD	1.6 pg	43
(CH$_3$)$_2$Sn^{2+}	(CH$_3$)$_2$SnH$_2$	AES	0.20 ng	43
		FPD	0.65 pg	43
(CH$_3$)$_3$Sn	(CH$_3$)$_3$SnH	AES	0.12 ng	43
		FPD	0.92 pg	43
Te	H$_2$Te	AA	11 ng	41

packing materials can cause serious losses of analytes when nanogram amounts are involved. Boiling with concentrated HCl or other severe treatment may be needed.

Cold trapping the hydrides results in the collection onto a small surface area of all of a particular metal hydride or organometallic compound in the scrubbed sample volume. Since the carrier gas volume containing the trapped compound removes it also in a small volume upon warming, a high degree of preconcentration is achieved. Trapped hydrides are carried out of the trap by a carrier gas in order of their boiling points and are generally well separated. Theoretical plate calculations have given plate numbers as high as 4600 for a 30-cm column. This effect is similar to that described by Kaiser (51), who developed a temperature gradient tube for cryogenic air preconcentration.

Cold trapping by absorbing hydrides (arsines) in cold toluene has been reported by Talmi and Bostick (see Figure 5 of Reference 49). Aliquots of the solvent were separated by gas chromatography and arsenic detected by microwave-stimulated atomic emission spectroscopy (AES). Although some loss of preconcentration is encountered because only a fraction of arsenic from each sample is analyzed, the excellent detection limit of the AES detector permits analysis of concentrations of arsenic compounds down to approximately 0.25 μg/L.

Detection of hydrides of the metals has been carried out by several different methods compared in Table 2. Electron capture detectors (ECD) lack specificity and vary considerably in response from one compound to another. Atomic absorption type detectors have some response problems because of the large active volume of the detector. Better response is being obtained more recently, as can be seen in the work of Andreae (46).

Flame ionization detectors (FID) respond with good sensitivity only to organometallic compounds. The hydrogen rich, hydrogen-air flame photometric detector (FPD) has proved to be a very favorable case for tin. Using a reversed flame burner (air inner cone), SnH emission bands are observed and well separated from the central flame cone. Detection limits have proved to be extraordinarily low, in the picogram range (48). The FPD has, of course, been widely used for sulfur in air analyses.

Atomic emission spectroscopic detectors (31, 47) have the best detection limits. Efforts have been made to directly couple such detectors to gas or liquid chromatographs. The use of various plasma emission spectroscopy detectors are described in several recent reviews (52, 53). These techniques include microwave-induced atmospheric-pressure plasma DC argon plasma and inductively coupled plasma spectroscopy. The microwave plasma has been widely used as a detector for gas chromatography.

There is an extensive literature of specific application of microwave plasma detection for gas chromatography that has been reviewed by Carnahan, Mulligan, and Caruso (53). The microwave plasma is particularly

TABLE 3. Microwave Plasma Detection Limits and Linear Dynamic Ranges[a]

Element + Emission Wavelength (nm)	Absolute Detection Limits (pg)	Detection Limits (pg/s)	Linear Dynamic Range
Microwave Plasma			
Hg(I) 253.7	60	0.60	10^3
B(I) 249.8	27	3.6	5×10^2
Al(I) 396.2	19	5.0	5×10^2
C(I) 247.9	12	2.7	10^3
Si(I) 251.6	18	9.3	5×10^2
Sn(I) 284.0	6.1	1.6	10^3
Pb(I) 283.3	0.71	0.17	10^3
P(I) 253.6	56	3.3	5×10^2
As(I) 228.8	155	6.5	5×10^2
Se(I) 204.0	62	5.3	10^3
S(I)[b] 545.4	140	52	5×10^2
Cl(I)[b] 479.5	310	86	5×10^2
Br(I)[b] 470.5	212	67	5×10^2
DC Plasma			
Cr[d] 267.7		4	
Cu[c] 324.7		5.6	
Ni[c] 341.4		320	
Si[d] 251.6		25	
Sn[d] 286.3		60	
Pb[d] 368.3		100	
Hg[c] 253.6		65	
B[d] 249.8		3	

[a] Reference 52.
[b] Values obtained with refractor plate spectral background correction.
[c] Values obtained for two-electrode plasma jet.
[d] Values obtained for three-electrode plasma jet.

well suited to be used with capillary columns since it is limited in sampling rate to a few milligrams per second. Typical detection limits are given in Table 3. DC plasmas have also been used as GC detectors and, as shown in Table 3, typically have poorer detection limits than the microwave plasma. However, the higher power levels of the DC plasma permit it to accept the much higher flow rates of packed columns.

The higher power levels also permit DC and inductively coupled plasmas to serve as detectors for high performance liquid chromatography (HCPL). Currently studies are focusing on the problems caused by non-aqueous solvents commonly employed (54–56) and it appears that HPLC with plasma detection will become an important technique for determining chemical speciation.

4. SELECTED SPECIFIC ELEMENTS

Environmental analyses for specific compounds have been carried out on an element-by-element basis as there is no realistic way to do otherwise. The following sections review what has been done in the case of several important elements. A salient feature of analytical importance in all this work is that analyte sample sizes are in the nanogram and sub-nanogram range. These low concentrations have in most instances required the use of preconcentration techniques and the employment of detectors having low limits of detection. Since preconcentration methods are used and small amounts of analyte are detected, a substantial need for specificity of detection arises to minimize possible interferences. The elemental specificity of emission-type detectors and their sensitivity has been used to good advantage. Unfortunately, many of the techniques employed are not really of a routine analytical nature. Carrying out chemical transformations at the nanogram level requires experience and skill to avoid mistakes. Consequently, most analytical methods must be considered to be research techniques. The trend in low analyte concentrations is not likely to change. If the study of tin can be used as an example, one should expect to have to analyze samples at the parts-per-trillion concentration level to observe methylated forms of the biomethylated metal elements.

In each of the following sections, presently identified forms of elements detected at trace or low concentrations in the environment are discussed.

Minerals of the elements have not been included. Elemental forms for some elements (i.e., Sb, As, Cu, Pb, and perhaps Sn) probably are present in trace amounts in nonmineralized water sediments. Nevertheless, their presence as a result of anaerobic reduction has not been demonstrated. This list of detected organometallic or organic compounds is deceptively short for elements such as As, S, Se, and Sn. Too many organic compounds of these elements are known and could exist under environmental conditions for the list provided to be considered complete.

4.1. Antimony

The chemistry of antimony and arsenic are quite similar in many ways. Consequently, one should reasonably expect to find $+3$ and $+5$ oxidation states and perhaps organometallic forms present in the environment. $RSbO(OH)_2$ (stibonic) and R_2SbOOH (stibenic) acids are known. Parris and Brinckman (57, 58) have studied the chemistry of methylantimony (and arsenic) compounds, which is important to environmental transport. Trimethylstibine was found more readily oxidized than trimethylarsine. If formed, it should be in the more water soluble form $(CH_3)_3SbO$ or its dimer. The methylantimony compounds may also be present as polymeric compounds. Despite the stability of the methylantimony compounds and the proven presence of methylarsenic compounds in the environment, no methylantimony compounds have as yet been detected in the environment. The biomethylation of antimony has apparently not been demonstrated even in laboratory work. Consequently, the definition of the antimony speciation problem is very much undecided.

Molecular forms of inorganic Sb(III) in aqueous media should be $Sb(OH)_2^+$, $Sb(OH)_4^-$, and $Sb(OH)_3$ in the pH 4–10 range according to Mesmer and Baes (59). Reduction potentials indicate that elemental antimony should exist under some conditions. However, little if any attention has been paid to separating the inorganic forms.

Nearly all environmental analyses have been total element analyses usually done by atomic absorption spectroscopy (AAS). Chambers and McClellan (60) were able to detect 10-ng/L Sb using conventional AAS and adding a solvent extraction concentrating step. The use of a graphite tube furnace AAS with antimony separated from the sample as SbH_3 has achieved a detection limit of 0.5 ng/L (61).

Talmi and Norvell (62) have determined antimony in environmental samples with detection limits of 0.05 using a microwave-stimulated atomic emission type detector. Antimony was converted to the triphenyl derivative by reaction with phenylmagnesium bromide and separated on a gas chromatographic column prior to detection.

The reduction of inorganic antimony to stibine and detection by atomic emission spectrometry has been developed (63) and applied to environmental analyses (48). This method is capable of detecting the methylantimony compounds, but none were found in any samples of air particulate or river, lake, and estuarine waters analyzed. Antimony was found to be an average of 75 ng/L in four samples of fresh water found to contain antimony and 27 ng/L in two saline water samples. The detection limit of the method was approximately 7 ng/L. In view of the fact that methyltin compounds were found at sub ng/L concentrations, a reexamination of antimony using a more sensitive detection system is in order. A review of the environmental chemistry of antimony is given in Figure 2.

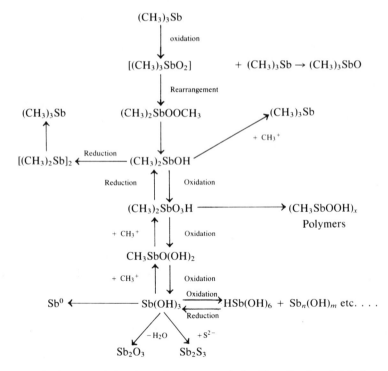

Figure 2. Environmental chemistry of antimony as derived from Parris and Brinckman (58) and Mesmer and Baes (59).

4.2. Arsenic

Speciation studies of arsenic have drawn considerable attention stemming from the toxicity of the element and the discovery in the last century that arsenic may be converted to a toxic airborne form. Early work by Challenger (64) is certainly classic in studying this biomethylation phenomenon. For excellent reviews on the subject of arsenic, the reader is referred to the National Academy of Science report (65) and the ACS Symposium Series No. 7, Arsenical Pesticides (66).

Arsenic analyses are classically carried out by the silver diethyldithiocarbamate spectrophotometric method (67) in which arsenic is converted to arsine by a zinc–hydrochloric acid reduction. The methyl and dimethylarsenic acid compounds are also reduced but do not give the same absorption curves. Antimony interferes because SbH_3 also gives a colored complex. Although suitable for speciation in the 10^{-6}-g As or greater sample size range, the only use made of this was by Peoples, Lakaso, and Lai (68), who analyzed bovine urine for inorganic and methylarsenic content.

Arsine is also generated by sodium borohydride reduction and this is widely used for total arsenic content using atomic absorption detection of the arsine (44). Sodium borohydride may also be used to reduce alkyl and arylarsenic compounds according to the chemistry in Table 4. A distinction may be made between arsenic(III) and arsenic(V). The former is reduced to arsine above pH 4, while the latter is not and requires pH 1.5 for complete reduction to arsine. Arsines generated in reaction chambers have been trapped in cold toluene and then analyzed by GC–AES as in the method of Talmi and Bostick (49) or cold trapped on a short column and separated by warming with AES detection in the method devised by Braman and Foreback (69), later described in detail and applied in analysis of natural water samples (31). Methyl–arsenic compounds have been found in a variety of natural waters in substantial percentages. Later work by Johnson and Braman (70) indicated that methyl–arsenic compounds were a small percentage of total arsenic in seawater but were associated with biota. Andreae (46), using a similar preconcentration and separation approach but an atomic absorption or electron capture detector, has studied arsenic speciation in seawater with similar results. Table 5 summarizes some of the available environmental data.

Arsenic speciation in air has been carried out using a glass wool filter to collect the particulate followed by silvered glass beads to trap volatile

TABLE 4. NaBH$_4$ Reduction Chemistry for Arsenic Compounds[a]

H$_3$AsO$_4$	pH 1.5	AsH$_3$
	pH > 4	No reaction
HAsO$_2$	pH 1.5	AsH$_3$
	pH > 4	AsH$_3$
CH$_3$AsO(OH)$_2$	pH 1.5	CH$_3$AsH$_2$
	pH > 4	Small amount prescribed
(CH$_3$)$_2$AsO(OH)	pH 1.5	(CH$_3$)$_2$AsH
	pH > 4	Major amount prescribed
(CH$_3$)$_3$AsO	pH 1.5	(CH$_3$)$_3$As
C$_2$H$_5$AsO(OH)$_2$	pH 1	C$_2$H$_5$AsH$_2$
C$_3$H$_7$AsO(OH)$_2$	pH 1	C$_3$H$_7$AsH$_2$
C$_4$H$_9$AsO(OH)$_2$	pH 1	C$_3$H$_7$AsH$_2$
C$_4$H$_9$AsO(OH)$_2$	pH 1	C$_4$H$_9$AsH$_2$
C$_6$H$_5$AsO(OH)$_2$	Ph 1.5	C$_6$H$_5$AsH$_2$
H$_2$N⎯⎯AsO(OH)$_2$	pH 1.5	H$_2$N, AsH$_2$

[a] From References 26 and 42.

arsines. The arsines are removed by mild NaOH wash and analyzed by the hydriding method (31). Most of the arsenic in ambient air appears to be in the inorganic form. Some methyl–arsenic compounds are found in the particulate and some trimethylarsine in the vapor form. Lawns treated with arsenic compounds rapidly produced methylarsines (29).

Substantial work has been done on speciation of arsenic in human urine. Peoples, Lakso, and Lai (68) first found indications that methyl–arsenic compounds were present in urine by their colorimetric method. Braman and Foreback (69) found a substantial amount of methyl–arsenic compounds in human urine. The work of Smith, Crecelius, and Reading (45) is particularly noteworthy. Using the hydride technique, they found that respired inorganic arsenic is not eliminated immediately but is converted principally to dimethylarsinic acid and also to methylarsonic acid. Elevated inorganic arsenic in air results in much increased dimethylarsinic acid in urine. Some data on speciation of arsenic in human urine from several sources are given in Table 6. A typical separation of arsines obtained in analysis of urine from a patient experiencing inhalation toxicity from arsenic is shown in Figure 3.

The hydriding analysis method for arsenic does have some limitations.

TABLE 5. Environmental Analysis for Arsenic (μg/L)

Location	As(III)	As(V)	Methylarsonic Acid	Dimethylarsinic Acid	Total
Fresh Water[a]					
Withlacooche River	<0.02	0.16	0.06	0.30	0.42
Remote pond	<0.02	0.32	0.12	0.62	1.06
Lake Carroll	0.89	0.49	0.22	0.15	1.75
Lake B	2.74	0.41	0.11	0.32	3.85
Saline Water[b]					
Tampa Bay	0.12	1.45	<0.02	0.20	1.77
Tidal flat	0.62	1.29	0.08	0.29	2.28
La Jolla, CA pier	0.034	1.70	0.019	0.12	1.87
Seawater, San Diego trough					
Surface	0.017	1.49	0.005	0.21	1.72
25 m	0.016	1.32	0.003	0.14	1.48
100 m	0.060	1.59	0.003	0.002	1.66
Rain[c]					
La Jolla, CA	<0.002	0.180	<0.002	0.024	0.204
	<0.002	0.094	<0.002	<0.002	0.094

Location	As(III) + As(V)	Dimethylarsinic Acid	Trimethylarsine
Air: Particulate and Vapor (ng/m³)[d]			
Suburban lawn	4.1	—0.90 (sum)—	
Suburban porch	0.5	1.0	0.4
Urban air	3.6	0.2	0
Urban air	0.5	trace	trace
Rural air	0.4	0.3	0
Greenhouse	1.7	0.4	20.5

[a] Reference 31.
[b] References 31 and 46.
[c] Reference 45.
[d] Reference 28.

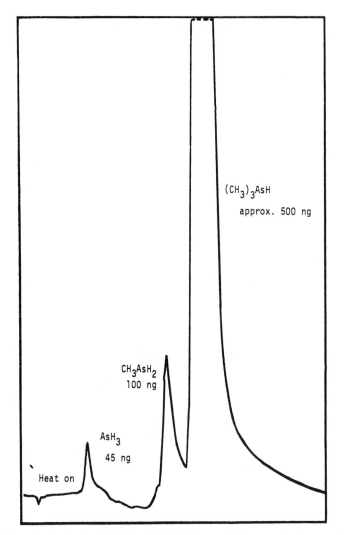

Figure 3. Analysis of human urine (2-mL sample) for arsenic compounds.

For example, it is not possible to determine the difference between $(CH_3)_3As$, $(CH_3)_3AsO$, and $(CH_3)_3As^{2+}$ (trimethylarsine itself will be a vapor form detectible without reduction). In addition, Penrose et al. (71) have observed in certain marine organisms, an organoarsenic compound not reduced by $NaBH_4$. Edmonds, Francesconi, and Cannon (72) have found this to be

$$
\begin{array}{c}
CH_3 \\
| \\
CH_3\text{———}As\text{———}CH_2COOH \\
| \\
CH_3
\end{array}
$$

This is probably the organoarsenic compound observed by Crecelius (73) to require digestion with 2M NaOH in order to decompose it and observe it as dimethylarsinic acid. Despite these difficulties, it is likely that the hydriding technique will remain the major approach for environmental arsenic speciation.

Speciation of arsenic in water and urine (39) and in biological samples (40) has been done by a method not requiring hydride reaction. Arsenic compounds in the presence of I^- ion are converted to iodides, which are then reacted with diethylammonium diethyldithiocarbamate to form the arsenic complexes of diethyldithiocarbomate.

$$As[(CH_2H_5)_2N\text{—}CS_2]_3$$

$$CH_3\text{—}As[(C_2H_5)_2N\text{—}CS_2]_2$$

$$(CH_3)_2\text{—}As[(C_2H_5)_2N\text{—}CS_2]$$

These compounds are separated by gas chromatography and detected using an electron-capture detector. Detection limits were 73, 40, and 15 ng/mL for inorganic arsenic, methylarsenic compounds, and dimethylar-

TABLE 6. Arsenic Speciation in Human Urine in μg/L (Average Values)

As(III)	1.3	1.9
As(V)	1.3	3.9
Methylarsonic acid	3.4	1.8
Dimethylarsinic acid	11.5	15.0
Total arsenic	21.2 ($\pm$ 2.04)	22.5 ($\pm$ 8.5)
Comments	Control group, $n = 41$	Not exposed, $n = 4$
	Reference 45	Reference 31

senic compounds, respectively. As discussed previously, the microwave plasma can be used as a specific element detector for arsenic (49, 62).

Analysis of elemental arsenic in the particulate or in anaerobic sediments has been largely neglected. This analytical problem needs solving since the methyl derivatives are not present in large amounts in the sediments. The environmental chemistry of arsenic is summarized in Figure 4.

4.3. Copper and Copper Complexes

4.3.1. Methods for Complexing Capacity

Determination of the complexing capacity of natural waters has been a subject of considerable controversy because of the somewhat empirical nature of the measurement. What is apparently desired is some measure of the ability of natural waters to detoxify metal ions contained therein through complexation with natural chelating agents. Each metal ion, of course, has its own set of formation constants with different complexing ligands and measurements. Therefore, using all metal ions of interest would give contradictory results. This problem is avoided by selecting copper(II) ion as the single test metal ion representative of all others with which to measure complexing ability. There are some problems. Copper(II) ion forms stronger complexes than do Ca^{2+} or Mg^{2+} ions with

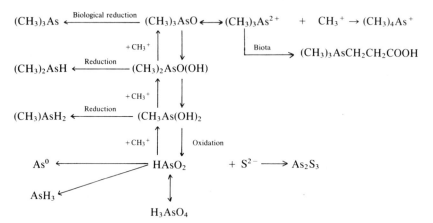

Figure 4. Environmental chemistry of arsenic as derived from Braman (29) and Mesmer and Baes (59).

naturally occurring ligands and thus would replace these ions in a complex. Therefore, measurements with copper do not give data on total free ligand concentrations.

Several approaches to analysis have been studied. Kunkel and Manahan (74, 75) developed a method based on the competition of natural ligands with OH^- ion for Cu^{2+} at pH 10. Copper(II) ion is added to a natural water sample, which is then adjusted to pH 10. After filtration to remove precipitated $Cu(OH)_2$, the copper remaining in solution is determined by atomic absorption analysis. Typical values for complexing capacity are in the 1–2 μmol/L range.

A column chromatographic method has also been reported. In this method (76), samples are spiked with an excess of copper(II) ion and passed through a column of Chelex 100. Copper strongly complexed passes through the column and is detected by AAS or AES. The method has been used in analysis of natural waters but may exhibit absorption effects or slow equilibration effects.

Some attempt has been made to relate complexing capacity of natural waters to molecular weight of the complexing agents. Smith (77) has determined the complexing capacity of fractions obtained by ultrafiltration using the copper titration method of Chau and Lum-Shue-Chan (78). This technique provides fractionation of the organic complexes into molecular weight ranges 100,000 and above, 50,000–100,000, 10,000–50,000, 1000–10,000, and 1000 and lower. Analyses were made of river water in an estuary location at the interference of the river with the Atlantic Ocean. Among the interesting results of the study was the fact that there appeared to be no correlation between the dissolved organic matter and complexing capacity (versus Cu^{2+}). This is quite in accord with the same result reported by others (80). Results also indicated that, while complexing capacity was distributed among the molecular weight fractions in fresh water, this was not so for the low salinity ocean–fresh interface waters analyzed. The large majority of complexing capacity was found in the 1000 or less molecular weight fraction.

Chau and Wong (79) have developed an ASV titrimetric method for the complexing capacity of natural waters. Water samples are titrated with copper(II) ions. The uncomplexed copper ion is detected after each addition of copper by ASV. This approach has gained reasonably wide acceptance.

Campbell et al. (80) have compared the copper(II) solubilization method (74), the zinc–zircon method (81), and the titration-ASV method (78). The zinc–zircon method gave markedly lower results than the others. These workers found also that copper may be solubilized as a colloidal complex in the copper(II) ion solubility method rather than as a true solution complex.

Two groups have detailed procedures whereby molecular weight information and formation constants may be obtained from ASV (82) and copper ion selective electrode measurements alone (83) during the titration method with copper(II) ion. The later paper contains a particularly elegant rendition of theory. The method was used to analyze a few natural waters. Molecular weights of ligands found were on the order of 1000.

Unraveling the speciation of metal complexes in natural water systems certainly will be aided by studies of the ligands present. It may in fact be far easier to determine what ligands are present and to calculate the distribution of complexed metal ion forms than to approach the problem by direct analysis. The factor of colloid formation appears to be important. What is surprising to this reviewer is the lack of information on sulfur-containing natural ligands, particularly sulfur amino acids or their disulfide oxidation products. A direct analysis approach that certainly appears to hold promise is the separation of preconcentrated ligands by HPLC. Applications will undoubtedly be developed as HPLC continues to gain popularity.

Considerable effort has gone into the determination of the structure of humic substances, a group of organic macromolecules found in natural waters and in soil. These compounds are generally considered to have complexing functional groups such as carboxyl, phenol, hydroxyl, and keto groups either isolated and acting alone or possibly acting as bidentate ligands in complex formation. Analysis of humic substances is done on large amounts of material rather than on traces found in natural waters. Humic substances are usually separated into humic acids, fulvic acids, and humin based on their solubility in acidic or basic solutions. These are then often considered to be defined materials for experimental purposes without further separation. Efforts are now being made to further define the humic substances after the initial separation. Methylation procedures (84–86) followed by separation on appropriate ligand chromatographic columns appear to hold good promise for identifying and perhaps isolating some of the compounds present. Two books on humic substances are available (87, 88).

4.3.2. *Environmental Complexes of Copper*

Considerable attention has been given to environmental studies of copper largely because of its toxic effects on fish and other biota. Although it has no reported stable organometallic compounds under environmental conditions, differences in toxicity to biota have been observed between "strong" and "weak" complexes of copper. There has been much activity in studies of this type. The effects of EDTA (89), NTA (90, 91), humic substances, and other materials, including inorganic complexes (92–94), on copper toxicity have been reported. In general, the conclusion is that toxicity of copper to fish follows the copper(II) ion activity as measured by ion selective electrode (95). Copper(II) ion activity is, of course, a function of the formation constant of copper complexes and ligand concentrations. None of the formation constants of the common inorganic form of copper in solution are particularly large; they are in the 10^6–10^{10} range. Phase diagrams from the review on cation hydrolysis by Mesmer and Baes (59) indicate that CuO should predominate even in natural systems containing carbonate while the hydrated cupric ion and $CuOH^+$ are major complexes found in solutions not having substantial amounts of the carbonate ion present.

The presence of naturally occurring organic complexing agents complicates the picture. Substantial competition for copper(II) ions must exist because of the well-documented decrease in copper ion activity in the presence of humic substances and fulvic acids (94). It appears that the use of a copper ion selective electrode to measure copper ion activity will be a better indicator of copper toxicity than a total copper analysis by, for example, AAS. Unfortunately, ion selective electrodes can take very long times to come to equilibrium especially at concentrations below 10^{-6} M. Blaedel and Dinwiddie (96) amply demonstrated this in their study of a micro flowthrough copper ion selective electrode. Response of the electrode was Nernstian in the 10^{-3}–10^{-8} M Cu^{2+} range but at best 6 h were required for steady state conditions for 10^{-8} M Cu^{2+} (0.6 ppb). The direct analysis of copper complexes after separation has apparently only been done for mixtures of NTA and EDTA complexes.

4.4. Germanium

The environmental chemistry of germanium is quite similar to that of boron. It usually appears in the environment as germanic acid, $Ge(OH)_4$

(pKa 8.59) or as salts of germanic acid. Like boric acid, it forms complexes with diols and polyols and thus may be found associated with or, to an extent, concentrated into organic matter.

Methylgermanium compounds such as methylgermanium trichloride, methylgermanium dichloride, trimethylgermanium chloride, and tetramethylgermane are stable with respect to the C—Ge bond and can exist as stable forms under environmental conditions, although likely as hydrated or hydroxylated forms of the corresponding methyl–germanium cation.

Only two studies that involve environmental speciation of germanium have been reported. Utilizing a hydride generation method similar to that for arsenic (31), germanium has been determined in seawater and in sargassum weed and its epiphytes (70). Inorganic germanium was found to be 0.042 ppb in seawater but concentrated by a factor of approximately 100 in the sargassum weed and epiphytes. An improved procedure (48) has been used to analyze fresh water, rivers, lakes, and saline bay waters in and near the Tampa, Florida, area. Inorganic germanium concentrations were found to be influenced by the proximity of coal-fired power generation plants. Water analyses ranged from 9–220 ppt Ge with the higher values observed near coal-fired power plants. Air analyses averaged 0.25 ng Ge/m^3.

The analytical methods used in these latter studies involved NaBH$_4$ reduction of germanium to germane under conditions at which all of the methyl–germanium compounds also would be reduced to corresponding hydrides (i.e., GeH$_4$, CH$_3$GeH$_3$, etc.) Despite the fact that detection limits for methyl–germanium compounds were near 4 ppt, no indication of their presence was observed in any of the water samples analyzed. Methyl–germanium compounds were also not found in air (48).

Future work with germanium should be to improve the limit of detection of the detection system and reexamine the analysis problem. The analysis of air for tetramethylgermane also needs to be studied since it is physically and chemically quite different from the other germanes. Barring new results to the contrary, germanium appears to exist in the environment simply as germanic acid or its complexes.

4.5. Lead

Most analytical work with lead has been of the total elemental concentration type. This is quite adequate for many uses, for example, medical clinical screening. Nevertheless, the stability of tetramethyllead, trimethyllead ion and dimethyllead ions in aqueous media make lead a good subject for speciation study.

Lead in aqueous media in the environment is most likely to be in the $+2$ oxidation state. Mesmer and Baes (59) have reviewed the hydrolysis chemistry of Pb^{2+} and methyllead cations. One should expect to find lead as $Pb(OH)_2$, $Pb(OH)_3^-$ when lead concentrations are low (the usual case), although polymeric lead ions such as $Pb_2(OH)^{3+}$ and $Pb_4(OH)_4^{4+}$ may be present. Insoluble common compounds likely to be present are PbO, $PbCO_3$, and $PbSO_4$.

Petrie and Baier (36) in their study of lead in organic free seawater by ASV and cyclic voltammetry found $PbOH^+$ to be a major fraction of lead near seawater pH values. At pH 8.5, the distribution was: $PbOH^+$, 88%; $PbCO_3$, 10%; $PbCl^+$ + Pb^{2+} + $PbSO_4$, 2%.

Although lead is reported apparently not to be methylated by methylcobalamine (97), tetramethyllead generation in lake water sediments has been demonstrated by Wong, Chau, and Luxon (99). This latter finding challenges the hypothesis that methylcobalamin is responsible for all environmental biomethylation. Nevertheless, a problem that it raises is: How is the monomethyllead (ion or complex) stabilized so that further biomethylation may occur? Dimethyllead and trimethyllead ions are reasonably stable in aqueous media as is tetramethyllead. The hydrolysis of dimethyllead cation is similar to that of the corresponding tin compound (59), the major forms are $(CH_3)_2Pb^{2+}$, $(CH_3)_2PbOH^+$, and $(CH_3)Pb(OH)_2$ in the pH 6–9 range at low lead concentrations.

No analytical methods applied to methyllead determination in nonlaboratory environmental samples have been reported. Laboratory study analyses for methyllead compounds (98) were carried out by gas chromatography after cold trapping gas samples as a means of preconcentration. It is apparent that a good approach to analysis of aqueous solutions for lead would be to attempt the covalent hydriding technique. Both Pb^{4+} type ions (but not Pb^{2+}) and $(CH_3)_3Pb^+$ will probably be converted to corresponding hydrides. A substantial decrease in the lower limit of detection of emission detectors would be necessary if methyllead compounds are very low in concentration as is the case for Sn. We have noted that precipitation of PbH_2 is likely upon reaction of Pb^{2+} with $NaBH_4$.

Airborne lead compounds are found in both gaseous and particulate matter. Most of the reported methods are incapable of distinguishing between gaseous material and those particles not retained by a filter. A comprehensive review of methods for determining nonfilterable lead in ambient air was given by Harrison and Perry (99). There is also often a problem of insufficient sensitivity except at heavily polluted sites. Recently a procedure has been reported (100) that permits detection of molecular lead as an organic compound in the gas phase by modifying the method of Harrison et al. (101). In this technique the gaseous lead alkyls are collected in an adsorption tube and then desorbed for atomic absorption spectrometric determination. In addition, a gas chromatographic separation into tetraethyl, tetramethyl, and mixed ethyl–methyllead compounds was possible for samples containing greater than 0.1 μg of organic lead per m^3 (101).

Lead compounds in soil were identified by x-ray powder diffraction (XRD) after separation by density gradient and magnetic fractionation by Olson and Skogerboe (102). The major compounds found were $PbSO_4$, PbO, and PbS. Elemental lead was found in one sample. The density gradient method was performed by centrifuging 10-g soil samples first in CCl_4 and then in mixtures of CCl_4 with CH_2I_2 of increasingly density. Nearly 80% of the lead was found associated with a comparatively high density (>3.32 g/cm^3) soil fraction obtainable by centrifuging with pure CH_2I_2. Since major soil components have a density lower than 3.32 g/cm^3, this treatment served as a good preconcentration step. In a similar study by Biggins and Harrison (103), lead compounds were separated from street dusts by magnetic and density gradient separation. Using XRD, they identified $PbSO_4$, Pb^0, $PbSO_4 \cdot (NH_4)_2SO_4$, Pb_3O_4, $PbO \cdot PbSO_4$, and $2PbCO_3 \cdot Pb(OH)_2$, although they conclude that the majority of lead does not exist in a crystalline form susceptible to XRD analysis.

Several general reviews of the environmental chemistry of lead are available. One is the report of a multiuniversity study of lead (104) and the other is a compilation of reports on lead by the U.S. Geological Survey (105). A review of lead in the human environment has recently been compiled by a National Research Council committee (106).

4.6. Mercury

4.6.1. Background

Current interest in speciation of all metals probably was initiated by studies on organomercury compound pollution and toxicity problems that came into prominence in the late 1960s. A large amount of research has since appeared on biotransformations of mercury, its aquatic chemistry, toxicity to biota, and monitoring and analysis techniques. The subject of the environmental chemistry of mercury has figured prominently at many scientific meetings. The following are a number of suggested references for review. L. J. Goldwater (107) has published an interesting historical account of mercury. The excellent   report of the Swedish expert group in 1971, *Methyl Mercury in Fish* (108), was certainly a catalyst for the work that has followed. Baughman et al. (109) did an outstanding study of the aquatic chemistry of organomercury compounds including studies on equilibrium reactions, photochemistry, acidolysis and modeling of aqueous equilibrium composition. An excellent review of the chromatographic and biological aspects of organomercurials was published by Fishbein (110) in 1970. Heaton and Laitinen (111) have reported an electrochemical study of methylmercury(II) ion in aqueous solution. Rabenstein et al. (112) have studied the hydrolysis equilibria of methylmercury cation in aqueous media. A review of the environmental chemistry of mercury, its pathways for exposure to man, and its toxicological effects has been prepared by a committee of the National Research Council (113).

Interest in the speciation problem in trace mercury analysis stems from the marked difference in toxicity between CH_3Hg^+ compounds, $(CH_3)_2Hg$, Hg^{2+} compounds, and elemental mercury. The methylmercury compounds have been implicated in a number of poisoning episodes (108) and thus much interest is attached to the determination of methylmercury compounds, especially in biological materials. The major analytical approach in speciation has been to differentiate between Hg (elemental), Hg(II) ion, CH_3Hg^+ ion, and dimethyl mercury with no consideration given to the determination of aqueous complex ions or of mercurous compounds. A summary of the environmental chemistry of mercury is given in Figure 5.

4.6.2.   *Water and Biological Materials*

Water and biological samples are usually extracted to remove mercury compounds and the organic extracting solvent is then analyzed by gas chromatography using electron-capture detectors. Electron-capture detectors have detection limits of approximately 1.0 ng per Hg sample and therefore are reasonably sensitive and probably adequate for most analysis purposes. Atomic absorption of elemental mercury after reduction

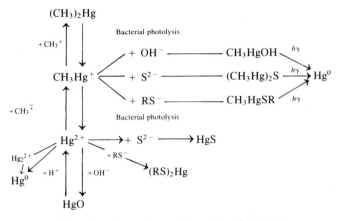

Figure 5. Environmental chemistry of mercury.

of samples is adequately sensitive and commonly used in many procedures for total mercury in water. These methods will not be discussed here.

Gas chromatographic methods have been widely used for organomercury determinations in water and in biological samples. Early procedures developed by Westoo (114, 115) were largely applied for analysis of biological materials. Methylmercury type compounds were converted to CH_3HgCl, extracted into organic solvents, and analyzed using an electron-capture detector. This approach to determination of organomercurials has the advantage that several organomercurials can be separated and individually identified if necessary although methylmercury compounds are usually the only types detected.

Most procedures for biological materials involve homogenizing biological samples, treatment with HCl, and extraction into an organic solvent such as benzene. Quantitative extraction of methylmercury chloride and clean up of the extract to eliminate co-extracted material from samples are the main problems. Extraction efficiency is certainly influenced by complexation of methylmercury cation by the sulfur amino acids in proteins and possibly also by inorganic sulfides.

Formation constants for some important mercury(II) and methylmercury(II) cations from Baughman et al. (109) are given in Table 7. Note that the sulfur complexes are much stronger than the inorganic ligand complexes. Some of the sulfur complexes may be of sufficient strength to resist dissolution even in mildly acidic media. The real strength of a protein-bound methylmercury cation may be somewhat greater than the formation constants indicate. Helical protein structures certainly inhibit removal of bound methylmercury or mercury (or other metal cations for that matter). To aid in recovery, Matsunaga et al. (116) added copper(II) chloride to fish homogenates. Copper appears to displace mercury from the protein amino acids. The production of copper metal in a sample solution upon reduction could be a problem with this approach. Cappon and Smith (117) have found that addition of urea to samples helps uncoil the amino acids and thus aids in recovery of methylmercury compounds. Strong reducing agents appear to be capable of reducing thiol complexed mercury. Toffaletti and Savory (118) found that $NaBH_4$ reduced both methyl and phenylmercury(II) cations even in the presence of cysteine. This writer has found that $NaBH_4$ will also reduce the dithizone complexes of methylmercury and mercury(II) ions to elemental mercury.

The determination of both inorganic and methylmercury in samples has always been somewhat of a problem. An early approach was to first determine methylmercury chloride by gas chromatography, determine total mercury after sample oxidation by flameless AAS, and assume that inorganic mercury is the difference between the two numbers. Both photochemical oxidation (119) and oxidation with ordinary oxidizing reagents (120) have been used. Imprecision in the results of each method of course adversely influences the resulting inorganic mercury numbers.

A similar possible approach is to designate inorganic mercury as that

TABLE 7. Formation Constants for Selected Mercury Complexes

Compound	log K_f
CH_3HgCl	5.45
CH_3HgOH	9.37
$(CH_3Hg)_2S$	21.2
$CH_3HgSCH_2CH_2OH$	16.12
$CH_3Hg \cdot SR$ (cysteine)	15.7
$(CH_3Hg)_2S_2O_3$	10.9
CH_3HgNH_3	7.6
$HgCl_2$	13.2
$Hg(OH)_2$	21.7
$Hg(SR)_2$ (cysteine)	41.0
HgS (pK_{sp})	51.0

fraction of the total mercury that is reduced by a mild reducing agent such as Sn^{2+} ion. Organomercury compounds are then determined as the difference between this number and total mercury found when a strong reducing agent such as $NaBH_4$ is used. A difficulty not usually considered in these methods is interference from inorganic sulfide ions. It appears unlikely that Sn^{2+} ion is capable of reducing HgS, for example, to get a good analysis value for inorganic mercury. The total versus inorganic mercury analysis of water and wastewater samples has been automated using an oxidation for total mercury. Samples in the 0.05 to 6 ng/L range could be analyzed (121). A column chromatographic separation has been studied. Isothiocyanatopentaaquochromium(III) ion forms polynuclear species with methylmercury, mercuric acid, and mercurous cations. Separation was obtained using a cation exchange column (122).

Arah and McDuffie (123) have used the UV absorption of $HgCl_4{}^{2-}$ ion, $3 \times 10^4 \ M^{-1} \ cm^{-1}$ at 230 nm for its determination in the presence of $CH_3HgCl_2{}^-$ ions. The methods does not have sensitivity sufficient for trace analysis but might be useful in laboratory studies if 5- or 10-cm cells are used.

The determination of dimethylmercury is less of a problem because it apparently does not form complexes with the organosulfur compounds. It does not, for example, form complexes with dithizone. Therefore, extraction of dimethyl mercury is more readily accomplished than for methylmercury compounds. Dimethylmercury does react with mercuric chloride.

$$HgCl_2 \ + \ (CH_3)_2 \ \rightleftarrows \ 2CH_3HgCl$$

This has occasionally been used in its determination (124). Dimethylmercury also undergoes hydrolysis. In 0.1 M HCl at 40°C it has a half-life of approximately 200 min (109).

An additional problem not generally commented on is the volatility of $HgCl_2$ itself. Although a solid, it has a 1-mm vapor pressure at 136°C and appears to sublime at a reasonable rate when near 100°C. This writer has had difficulty in separating it from the supposedly more volatile methylmercury chloride on occasion.

More recently, derivatization approaches have been used prior to gas chromatography to get the mercury compounds present into a more readily analyzed form. Alkyl or aryl mercury compounds are formed by one of a variety of alkylating reagents followed by extraction and gas chromatography.

Jones and Nickless (125) made use of the Peters reaction (126) to prepare aryl mercury chloride compounds. Mushak et al. (127) later prepared pentafluorobenzene

$$ArSO_2H + HgCl_2 \rightarrow ArHgCl + SO_2 + HCl$$

mercury compounds by the same type reaction to obtain better sensitivity when using an electron-capture detector in the GC analysis step.

Several different alkylating reagents have been studied by Zarnegar and Mushack (128) including the types $[Co(III)(CN)_5R]^{3-}$, R_4Sn, $[ClCH_2Cr(III)(H_2O)_5]^{2+}$, $R_2Tl(III)(O_2CR^1)$, and $(Ar_4B)^-$. Alkylation efficiencies were less than quantitative and depended on the compound. Trimethylsilyl derivatives (129) have also been prepared.

Perhaps the most carefully detailed and successful recent work reported using alkylation for analysis of a variety of biological materials is that of Cappon and Smith (117). This method, as is the case of most others in biological sample analysis, is necessarily long and involved but includes use of a mercury-203 isotope tracer to provide a correction for recovery in each analysis and thereby improve accuracy.

Derivitization appears to be a good approach for facilitating the separation of mercury from sample materials. If all mercury compounds can be converted to the dialkyl forms without losing their original speciation, separating from sample materials may be facilitated and analyses shortened.

Some typical data on environmental analyses for mercury in water and biological samples are given in Table 8. It is apparent from these data that ambient mercury speciation work will continue to require a high degree of preconcentration in order to obtain sufficient sample for analysis. The speciation of mercury in seawater remains an unsolved analysis problem.

4.6.3. Air

The determination of mercury in air has been done by many different investigators over a long period of time. Many references can be found on different types of sampling and analysis instruments. Almost all measure the ultraviolet absorption of elemental mercury at 253.7 nm. Con-

TABLE 8. Mercury in Water and Biological Materials

Sample	Results	Reference
Tampa Bay, saline water		
(Nine different locations)		
8/18/71	0.13($\pm$0.04) μg/L Hg total	63
9/15/71	0.28($\pm$0.015) Hg total	
11/10/71	0.96($\pm$0.60) Hg total (approximately 50% CH_3H_5Cl)	
Sargasso Sea water		
3–100 m	0.01 μg/L, $n = 7$	70
Sea of Japan	0.15 μg/L	109
Human blood samples		
Clotted cells	71 μg/L, $n = 19$	
	Range 0.05–406	
Whole blood	6.7 μg/L, $n = 59$	
	Range 0.05–23	
Human urine	Inorganic Hg 4.86 μg/L, range 1.7–9.15	
	Organic 0.93, range 0.04–3.07, $n = 4$	
Marine fish (Norway)	44–155 ng/g	109
Fresh water	0.1 μg/L	109
Freshwater Fish	76–167 μg/g	109
Osprey feathers	17 μg/g	109
Pork chop	6–16 ng/g	109
Egg white	6–27 ng/g	109

centrations as low as 15 ng/m^3 may be detected (130); even lower concentrations may be found if preconcentration by amalgamation is used with the UV absorption methods. These methods are satisfactory for many monitoring purposes since mercury in polluted locations is almost exclusively in the elemental form. Nevertheless, these methods do not detect any organomercury or inorganic mercury compounds.

The speciation of mercury in air cannot be accomplished as easily as the determination of elemental mercury. The total mercury in air in nonpolluted locations out-of-doors is on the order of 3–6 ng Hg/m^3. The task of dividing this amount up among several types of compounds is a formidable one. Three reported studies have dealt with the speciation of mercury in air. All are based on the specific absorption of mercury compounds. Braman and Johnson (26) developed a train of absorption tubes (Figure 6) which, after removal of airborne particulate matter, isolate $HgCl_2$ type compounds, CH_3HgCl type compounds, elemental mercury, and dimethylmercury from each other. Elemental mercury absorbs onto silver-metal-coated glass beads while dimethylmercury does not but is

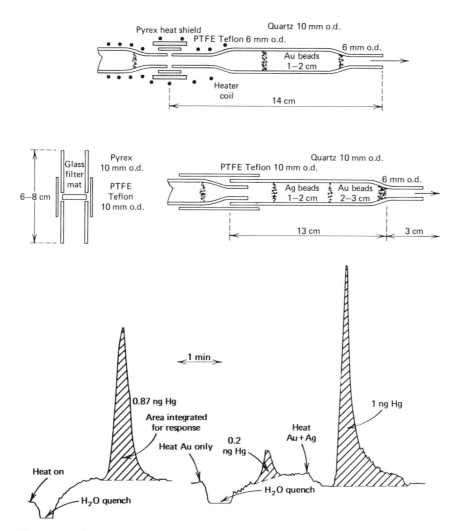

Figure 6. Adsorption tube stack design, sample transfer tube, and typical response recordings for mercury analysis. Reprinted with permission from R. S. Braman and D. L. Johnson, *Environ. Sci. Technol.* **8**, 996 (1974). Copyright 1974 American Chemical Society.

absorbed onto gold-coated glass beads. The $HgCl_2$ type compounds and CH_3HgCl type compounds are absorbed onto gas chromatographic solid supports treated with HCl and NaOH, respectively, to make use of the complex $HgCl_4^{2-}$ in one instance and the low volatility of $(CH_3Hg)_2O$ in the other. This method was used in a number of environmental studies (131, 132). The work was repeated by Rogers (133), who found similar results. Most of the mercury found in the environment was elemental mercury but a substantial fraction occurred in methylmercury forms. Only a very small percentage of mercury was found in the particulate.

Typical results on mercury in air speciation from a variety of sources are given in Table 9. Several total mercury in air determinations were

TABLE 9.   Mercury in Ambient Air in ng/m^3 [a]

Location	Hg Particulate	$HgCl_2$	CH_3HgCl	Hg^0	$(CH_3)_2Hg$	Total
Urban	12	0	3	4	1	20
Suburban	2	22	14	4	0	42
Bayside	3	10	54	6	3	76
Sargasso Sea	0.1	10	0.2	0.9	0	11
Sargasso Sea	3.1	0	0.7	1.1	0	4.7
Average of 19 daytime samples (Suburban)	0.27	0.86	0.63	2.67	0.05	4.48
Average of 13 nighttime samples (Suburban)	0.17	1.58	1.56	5.03	0.06	8.40
Classroom (before class)	2.9	4.2	12	310	0	367
Classroom (after class)	0	12	25	500	12	549
House						
Bathroom	—	43	4.4	200	0	247
Kitchen	—	12	0	88	0	100
Bedroom	—	0	10	37	0	47
Living room	—	2.7	23	110	0	136

[a] References 132 and 133.

done in addition to the speciation work referenced in Table 9. Air out-of-doors averaged 3–6 ng/m^3. Air in buildings was on the order of 100 ng/m^3. The source of this mercury indoors, mostly elemental mercury, is not known with certainty.

The detector used with this separation technique was a helium dc discharge emission (AES) type detector that had a detection limit near 0.01 ng Hg. Thus, with preconcentration of 50–200 L of air, concentrations of mercury down to a fraction of a nanogram per cubic meter was detected. Total analyte sample sizes were generally small, even with the preconcentration 0.1–10 ng.

Trujillo and Campbell (134) have reported a similar method for mercury speciation. Their absorption train consists of Carbosieve B, which absorbs both methylmercury type compounds and dimethylmercury, and a silvered chromatographic support. Detection of desorbed mercury was done by the UV absorption method. This method did not differentiate between $HgCl_2$, CH_3HgCl, and $(CH_3)_2Hg$.

All of the preceding absorption methods should be improvable. Particularly useful would be a specific absorption method that would permit release and detection of $HgCl_2$ and CH_3HgCl as the compounds absorbed. In the development of the technique noted earlier (26), this writer found that elimination of background mercury from absorption materials was a very difficult problem. Only by careful control of absorption tubes could good data be obtained. Metal-coated surfaces were easier to maintain clean than coated gas chromatographic packing materials. Developing a selective metal surface for absorption followed by perhaps a chemical displacement desorption would be one approach to study.

The collection of all mercury compounds on a solid sorbent is also a feasible approach, as has been demonstrated (26). If this could be done followed by a separation of sorbed compounds and detection by some emission type technique, the desired speciation method would be improved. Adaptation of this approach to analysis of samples in the sub-nanogram range will still prove difficult largely because of laboratory contamination problems. Most buildings, even ones not containing laboratories, have mercury in air concentrations on the order of 100 ng/m^3. Most laboratories have concentrations well above this, probably in the range 400–800 ng/m^3. It is virtually impossible to handle mercury sampling tubes for ambient air analyses at 3–6 ng/m^3 in a laboratory at 400 ng/m^3. Elimination of interferences requires that work be done in a nonlaboratory room, a precautionary measure taken in the development and use of the mercury speciation method (26).

As is the case for water analyses, mercury speciation in air requires a high degree of preconcentration prior to detection by a suitable AES detector. No other detector as yet developed for mercury detection has the detection limits of the AES detectors.

4.7. Sulfur

Environmental analyses for this element have historically been treated quite differently from those for the metals reviewed in this chapter. The determination of sulfur is more often carried out as a determination of a specific sulfur compound than as a determination of total element composition. Literature on the determination of specific sulfur compounds is extensive. Many methods exist for the determination of sulfate ion or sulfur compounds at parts-per-million concentrations or of sulfur dioxide in air. This discussion of speciation of sulfur will be limited to certain specific analysis problems: the determination of trace-reduced forms of sulfur in water and in air in the vapor state.

Although the toxicity of H_2S is known to be considerable, not a great deal of current work is done in toxicity studies. It is important then to note the work of Smith et al. (135) on the toxicity of H_2S to fathead minnows. Concentrations of hydrogen sulfide in water in the parts-per-billion range showed toxic effects.

Pulse polarography (136) also has good sensitivity for sulfide ion and improved selectivity. Sulfide ion was determined down to $3.5 \times 10^{-7} M$. Renard et al. (137) found that sulfide ion and mercaptans could be distinguished from each other as could thiosulfite, sulfide, and sulfite ions.

Analyses of samples for sulfur compounds are usually done by the highly selective and sensitive FPD. This detector can be either used alone without separation or with a gas chromatographic column or selective scrubber prior to detection. The detector originally developed by Brody and Chaney (138) observes the S_2 band emission produced when sulfur compounds are introduced into a hydrogen-rich, hydrogen–air flame. Since organic compounds in general produce comparatively weak emission bands in this type of flame, selectivity is high in favor of sulfur. Sensitivity of the detector for sulfur is outstanding: The limits of detection are near 0.01 ng S per sample when the FPD is used as a gas chromatographic detector.

The response of the FPD to sulfur compounds is not linear; it is a squared function of the amount of sulfur present in the flame. Greer and Bydalek (139) found that response obeys the equation:

$$R = k\ k'K\quad S^2(10^{-\alpha K S^2})$$

where detector response, R, in arbitrary units is a function of sulfur mass S, the constants k, k', K, and α are determined experimentally. The equation corrects for self-absorption, which is observed to be important in samples larger than 60 ng S.

The sulfur FPD detector was developed by Stevens et al. (140) into a gas chromatographic air analyzer. The light molecular weight sulfur compounds were separated on a polyphenyl ether coated Teflon chromatographic column, which minimized losses of sulfur compounds. Limits of detection for the analysis of 10 mL air samples were approximately 0.2 ppb (volume). In use this method has proven to be best applied to urban air analyses.

Ambient concentrations of total sulfur in air removed from proximity of pollution sources are well below 1 ppb. Unless preconcentration methods are used, even the highly sensitive FPD–GC system is not sufficiently sensitive for analyses. Only limited work has been done on speciation below 1 ppb. Natusch et al. (24, 141) developed an impregnated filter method for preconcentrating H_2S. They found H_2S was present in the range 5–10 ppt in air (142).

Braman and Ammons (30) have studied the preconcentration of H_2S and other light sulfur compounds on gold-coated glass bead surfaces. Sulfur compounds are desorbed by H_2 while heating the tube and are detected by means of a FPD. Hydrogen sulfide and dimethylsulfide are desorbed as the compounds absorbed. The preconcentration technique permitted determination of sulfur compounds at the parts-per-trillion concentration range. Analyses of ambient air in the presence of plants indicated that a mixture of unidentified organic sulfur compounds may make up a substantial fraction of the total reduced forms of sulfur in air.

Recently, substantial improvements have been reported in the preconcentration and analyses of reduced sulfur compounds in the atmosphere. Using cryogenic enrichment sampling and wall-coated, open tubular, capillary column, cryogenic gas chromatography with an FPD detector (143), concentrations of H_2S, COS, dimethylsulfide (DMS), CS_2, dimethyl disulfide (DMDS), and methyl mercaptan (MeSH) were measured at a variety of locations across the eastern and southeastern United States (144, 145) with fluxes of total sulfur ranging from 0.001 to 1940 g $S/m^2 \cdot yr$. The flux values from a variety of environments are summarized in Table 10.

4.8. Selenium

Selenium is similar to sulfur in much of its chemistry; consequently, its environmental chemistry is similar. Oxidation states are -2, 0, $+4$, and $+6$. Heavy metal selenides are insoluble, as is elemental selenium. Elemental selenium is not rapidly oxidized and is only slowly available to plants from soils (146, 147). Probably the major forms of selenium in the aerobic environment and even in mildly anaerobic sediments or soil, it should be immobilized there for very long periods of time.

Hydrogen selenide oxidizes readily in the presence of oxygen to form elemental selenium. Thus any biological reductions to H_2Se should contribute to the production of elemental selenium. Selenous acid or its salts

TABLE 10.   Fluxes of Reduced Sulfur Compounds from Various Environments[a]

	Average Sulfur Flux ($g_S/m^2 \cdot yr$)					
	H_2S	COS	DMS	CS_2	DMDS	S
Intertidal marsh (Delaware)	0.096	0.006	0.042	0.013	0.0004	0.16
Adjacent marsh (infrequently flooded)	0.01	0.020	0.91	0.12	0.006	1.16
Cultivated histosol soil (northwestern New York)	0.168	0.027	0.005	0.127	0.004	0.331
Cultivated histosol soil growing potatoes	0.164	0.019	0.008	0.171	0.001	0.361
Stagnant freshwater swamp	0.166	0.005	0.004	0.006	—	0.182

[a] Reference 145.

are the major water soluble selenium species. Selenous acid is a weak acid, much more mobile in the aqueous environment and much more available to plants (147). Selenates are reasonably strong oxidizing agents and should be readily reducible to selenites under most environmental conditions. The selenites are more stable in alkaline media.

The simple methylselenium compounds have been found to be produced from natural sources. Methyl selenol, CH_3SeH, is the simplest compound but is apparently readily oxidized to CH_3—Se—Se—CH_3, which has been found in plants (148). Dimethylselenium, $(CH_3)_2Se$, is produced by microbiological activity (149, 150).

The major selenium metabolite exhaled by rats has been found to be dimethylselenium (151). Biomethylation of inorganic selenium is considered to be a detoxification mechanism since dimethylselenite is shown to be 1/500 as toxic as selenite compounds (152). Trimethylselenonium compounds, $(CH_3)_3Se^+$, have been found as a urinary form of selenium (153, 154), thus pointing to the possibility that this type of compound could be a prevalent form of selenium.

It has been known for a long time that a number of plants accumulate selenium to a great extent, some over 1000 ppm. The excellent review on selenium published by the National Research Council (155) should be consulted.

Several different selenium compounds have been reported to be present in plants, many of which are selenium analogs of the sulfur amino acids such as

Methylselenocysteine, CH_3—Se—CH_2—$CH(NH_2)COOH$ (156),
Selenomethionine, CH_3—Se—CH_2—CH_2—$CH(NH_2)COOH$ (157),
Selenomethylselenomethonine,
   $(CH_3)_2Se^+$—CH_2—CH_2—$CH(NH_2)COOH$ Cl^- (158),
Dimethylselenide (158),
Dimethyldiselenide (148).

Cleavage of some of these methylselenium amino acids could easily lead to the production of the volatile selenium compounds dimethylselenide and dimethyldiselenide and thus account for the observations of these compounds. It is also likely that the large losses of selenium from plant material during ashing procedures prior to analysis are attributable to decomposition of methylselenium amino acids, of which the few indicated previously are by no means a complete list. The environmental analysis problem for selenium is made a bit more complex by the fact that there are several reported selenium–sulfur compounds of the type R—S—Se—S—R (159) and R—S—SeH (160), although the latter is less stable. Figure 7 summarizes the environmental chemistry of selenium.

Analytical procedures used by Ganther (159, 160) for solution analysis for the amino acid type compounds were thin layer chromatography with ninhydrin developer, column chromatography, and electrophoresis. Metal chelate columns were used. Both copper and nickel columns gave good separations. Detection was done by spectrophotometry and by count rate in experiments in which selenium-75 isotope was used. Although the methods noted earlier were not really applied to trace analyses of the environment, the adaptation to this type of analysis appears possible with the addition of highly sensitive detectors or detection systems for selenium.

Quantitative analytical procedures for speciation of selenium in the environment are apparently totally lacking. Volatile selenium compounds from plants have been collected by cold trapping procedures, followed by gas chromatography, but these have not been applied to ambient air analyses. All other analyses have been for total selenium or are in fact

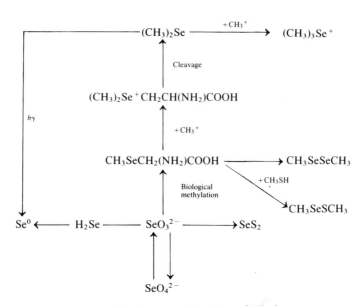

Figure 7.   Environmental chemistry of selenium.

**TABLE 11. Environmental Analyses for
Selenium (SeO_3^{2-})[a]**

Location	Se, µg/L
Palm River	17
Lake Carroll	<0.2
Hillsborough Bay	38
15th Fairway Pond	<0.2
Alafia River	<0.2
Gulf of Mexico	<0.2
Rain	<0.2

[a] Reference 164.

methods which, because of their chemistry, determine total selenium in solution, probably in the form of the selenite ion. A few environmental selenium analyses are given in Table 11 to indicate the sample size problem.

A few of the more sensitive selenium methods should be mentioned since combinations of these with appropriate separation techniques will eventually provide the speciation methods needed. Most trace selenium work has been done using H_2Se generation by $NaBH_4$ reduction followed by detection by AAS (44, 161). A flameless atomic absorption method has also been reported (162) as suitable for analyses of concentrations as low as a few parts per billion. The flameless atomic absorption method does have substantial interference problems from metallic cations, and an ion exchange treatment is needed.

Nelson (163) has developed a H_2Se evolution technique using reduction by Cr(II) ion in acidic media followed by detection of trapped H_2Se by AES. The method required a careful exclusion of oxygen to avoid losses of H_2Se. Detection limits were near 5 ng of selenium.

A sensitive gas chromatographic microwave-induced emission plasma method has been developed based on the preparation of a volatile selenium complex (50). This technique has a detection limit near 40 pg per sample and thus is superior in sensitivity to the atomic absorption methods.

In organic selenium analysis, the combination of AAS with gas chromatography has been used for analysis of dimethylselenide and dimethyldiselenide mixtures (164). Although the method has not been applied

to environmental analyses, it could readily be used for them with some
development of an appropriate preconcentration sampler.

4.9. Tin

The most widely found and most abundant compound of tin is SnO_2, quite
insoluble in water. The major tin ore, cassiterite, is SnO_2. Tin(II) ion can
exist in the environment under reducing conditions. It is readily oxidized
to Sn^{4+} compounds, which can form soluble ions and complexes of the
type $Sn(OH)^{3+}$, $Sn(OH)_2$, $Sn(OH)_3{}^+$, and $Sn(OH)_4$, probably with some
dimerization (59). Methyltin compounds and ions are stable in aqueous
media; the compound $(CH_3)_4Sn$ is comparatively nonreactive and volatile. Methyltin cations hydrolyze to form several different complex ions
with hydroxide ion such as $(CH_3)_2Sn(OH)_2$. Tin is one of the elements
that can be methylated by methylcobalamine, so one would expect to find
methyltin compounds in the environment. The environmental chemistry
of tin is summarized in Figure 8.

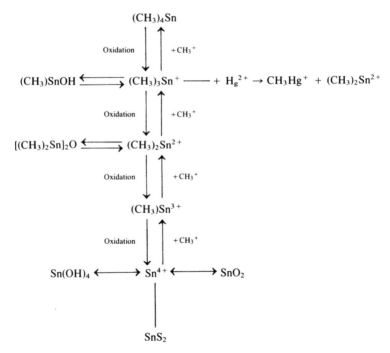

Figure 8.   Environmental chemistry of tin.

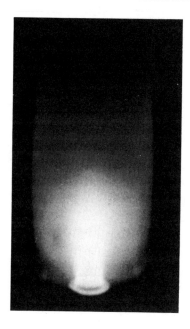

Figure 9. View of the SnH emission in a reversed FPD.

Environmental analyses for tin have generally been of the total elemental composition type. The NaBH$_4$ hydriding process followed by AAS has proved to be the most sensitive method but it has not been used in speciation studies. Braman and Tompkins (43) have developed a method for the analysis of aqueous solutions for the inorganic and methyltin compounds based on the NaBH$_4$ hydride generation approach used previously for arsenic (31). All tin compounds are reduced to corresponding hydrides in a buffer at pH 6.5 by reaction with NaBH$_4$. The hydrides are cold trapped at liquid nitrogen temperatures, separated upon warming, and detected using a reversed (air in the center cone, H$_2$ outside) hydrogen-rich, flame emission type detector (see Figure 9). This detector was similar in design to that used by Aue and Hill (165) for detection of organometallic compounds. The detector is based on the observation of SnH emission bands, which are produced in the presence of very little background noise, thus permitting the extremely low noise limited limits of detection, ~1 pg. Since 100 mL of sample may be analyzed the concentration detection limit is near 0.01 ppt.

Methyltin compounds were found in a variety of samples, for example, freshwater lakes and rivers, rainwater, and saline waters. Table 12 gives selected environmental analysis results. This confirms the suspected

TABLE 12.   Selected Environmental Analyses for Tin in ng/L[a]

Location	Tin(IV)	Methyl tin	Dimethyl tin	Trimethyl tin	Total
Lake Carroll	7.7	—	0.96	3.3	12
Lake Eckles	10	0.99	1.2	7.6	38
14th Fairway Pond USF	2.7	1.4	0.67	1.3	21
Withlecochee River	1.0	2.8	2.2	trace	6.1
Hillsborough River	0.41	0.49	0.36	trace	1.3
Average of 14	4.2	2.0	1.4	1.5	9.2
		Saline Estuarine Waters			
Gulf of Mexico	4.5	9.62	3.2	—	8.3
McKay Bay	20	trace	2.2	0.45	23
Old Tampa Bay	1.4	0.86	2.0	0.65	5.0
Palm River	567	—	4.6	4.0	576
Average of 7	.7	0.63	1.4	0.50	4.2
		Rain			
Average of 10	11	5.9	7.4	0.22	25

[a] Reference 43.

biomethylation process for tin, but concentrations obtained, well down in the parts-per-trillion range, also confirm the need for the highly sensitive detector developed.

A method was also developed for tetramethyl tin in air based upon the absorption of the compound onto Chromasorb 102. Air samples up to 30 L in volume could be preconcentrated. Traces of tetramethyltin were detected in air but an improvement in the preconcentration device is needed to obtain quantitative data. Nevertheless, the detection of methyltin compounds in rain, tetramethyltin in air and methyltin compounds in natural waters indicates that the element is extensively involved in a global biogeochemical cycle.

Tin(II) ion was not found in any of the samples analyzed. It can be determined as the difference between a sample treated with a small amount (1–2 drops) of a 0.01 M solution of iodine prior to analysis and a sample analyzed without the iodine treatment. Iodine oxidized Sn(II) to Sn(IV) under the conditions employed, but it does not oxidize methyltin compounds.

REFERENCES

1. R. K. Stevens and W. F. Herget, Eds., *Analytical Methods Applied to Air Pollution Measurements*, Ann Arbor Science Publishers, Ann Arbor, MI, 1974.

2. E. D. Hinkley, Ed., *Laser Monitoring of the Atmosphere*, Springer-Verlag, Heidelberg, 1976.

3. E. D. Hinkley, *Environ. Sci. Technol.* **11**, 564 (1977).

4. E. D. Hinkley and A. R. Calawa, Diode Lasers for Pollution Monitoring, in Reference 1, pp. 55–70.

5. J. N. Pitts, Jr., B. J. Finlayson-Pitts, and A. M. Winer, *Environ. Sci. Technol.* **11**, 568 (1977).

6. E. C. Tuazon, A. M. Winer, and J. N. Pitts, Jr., *Environ. Sci. Technol.* **15**, 1232 (1981).

7. B. D. Green and J. I. Steinfield, *Environ. Sci. Technol.* **10**, 1134 (1976).

8. A. P. Altshuller, *J. Gas Chromatogr.* **1**(7), 6 (1963).

9. A. P. Altshuller, W. A. Lonneman, F. D. Sutlerfield, and S. L. Kopezynski, *Environ. Sci. Technol.* **5**, 1009 (1971).

10. T. A. Bellar, M. F. Brown, and J. E. Sigsby, Jr., *Anal. Chem.* **35**, 1924 (1963).

11. W. A. Lonneman, T. A. Beller, and A. P. Altshuller, *Environ. Sci. Technol.* **2**, 1017 (1968).

12. E. D. Pellizzari, J. E. Bunch, B. H. Carpenter, and E. Sawicki, *Environ. Sci. Technol.* **9**, 552 (1975).

13. J. S. Parsons and S. Mitzner, *Environ. Sci. Technol.* **9**, 1053 (1975).

14. A. E. Dravnicks, B. K. Krotoszynski, J. Whitfield, A. O'Donnell, and T. Burgwald, *Environ. Sci. Technol.* **5**, 1220 (1971).

15. U.S. Environmental Protection Agency, *Sampling and Analytical Procedures for Screening of Industrial Effluents for Priority Pollutants*, EPA, April 1979.

16. National Institute of Occupational Safety and Health, *NIOSH Manual of Analytical Methods*, 2nd edition, Volume 3, 1979.

17. National Research Council Report, *Vapor-Phase Organic Pollutants*, National Academy of Sciences, Washington, D.C., 1976.

18. W. N. Billings and T. F. Bidleman, *Environ. Sci. Technol.* **14**, 679 (1980).

19. V. H. Regener, *J. Geophys. Res.* **65**, 3975 (1960).

20. V. H. Regener, *J. Geophys. Res.* **69**, 3795 (1964).

21. J. A. Hodgeson, K. J. Krost, A. E. O'Keefe, and R. K. Stevens, *Anal. Chem.* **42**, 1795 (1970).

22. J. D. Sensenbaugh and W. C. L. Hemeon, *Air Repair* **4**, 5 (1954).

23. A. F. Smith, D. G. Jenkins, and D. E. Cunningworth, *J. Appl. Chem.* **11**, 317 (1961).

24. D. F. S. Natusch, H. B. Klonis, H. D. Axelrod, R. J. Teck, and J. P. Lodge, Jr., *Anal. Chem.* **44**, 2067 (1972).

25. S. Hochheiser and L. A. Elfers, *Environ. Sci. Technol.* **4**, 672 (1970).

26. R. S. Braman and D. L. Johnson, *Environ. Sci. Technol.* **8**, 996 (1974).

27. M. A. Tompkins, Environmental Analytical Studies of Antimony, Germanium and Tin, Ph.D. thesis, University of South Florida, 1977.

28. D. L. Johnson and R. S. Braman, *Chemosphere*, **6**, 333 (1975).

29. R. S. Braman, Arsenic in the Environment in Arsenical Pesticides, ACS Symposium Series No. 7, American Chemical Society, 1975, pp. 108–123.

30. R. S. Braman, J. M. Ammons, and J. L. Bricker, *Annal. Chem.* **50**, 992 (1978).

31. R. S. Braman, D. L. Johnson, C. C. Foreback, J. M. Ammons, and J. L. Bricker, *Anal. Chem.* **49**, 621 (1977).

32. R. S. Braman and T. J. Shelley, Diffusional Discrimination and Reversible Chemisorption for Speciation in Air, in *Environmental Speciation and Monitoring Needs for Trace Metal-Containing Substances from Energy-Related Processes*, F. E. Brinckman and R. H. Fish, Eds., NBS Special Publication 618, U.S. Government Printing Office, Washington, D.C., 1981.

33. R. S. Braman, T. J. Shelley, and W. A. McClenny, Tungstic Acid for Preconcentration and Determination of Gaseous and Particulate Ammonia and Nitric Acid in Ambient Air, *Anal. Chem.* **54**, 358–364, (1982).

34. T. R. Copeland and R. K. Skogerboe, *Anal. Chem.* **46**, 1257A (1974).

35. Y. K. Chau and Lum-Shue-Chen, *Water Res.* **8**, 383 (1974).

36. L. M. Petrie and R. W. Baier, *Anal. Chem.* **50**, 351 (1978).

37. M. S. Frant, *Environ. Sci. Technol.* **8**, 224 (1974).

38. R. P. Buck, *Anal. Chem.* **48**, 23R (1976).

39. E. H. Daugherty, A. W. Fitchett, and P. Mushak, *Anal. Chem. Acta* **79**, 199 (1975).

40. P. Mushak, K. Dessauer, and E. L. Walls, *Environ. Health Perspectives* **19**, 5 (1977).

41. J. A. Fiorino, J. W. Jones, and S. G. Capar, *Anal. Chem.* **48**, 120 (1976).

42. G. Nelson, Analytical Method for Aqueous Selenite Ion in the Environment, M.S. thesis, Dept. of Chemistry, University of South Florida, Tampa, FL 1976.

43. R. S. Braman and M. A. Tompkins, *Anal. Chem.* **50**, 1088 (1978).

44. F. Fernandez, *At. Absorp. Newsl.* **12**, 93 (1973).

45. T. J. Smith, E. A. Crecelius, and J. C. Reading, *Environ. Health Perspectives* **19**, 89 (1977).

46. M. O. Andreae, *Anal. Chem.* **49**, 820 (1977).

47. C. Feldman and D. A. Batistoni, *Anal. Chem.* **49**, 2215 (1977).

48. R. S. Braman and M. A. Tompkins, *Anal. Chem.* **51**, 12 (1979).

49. Y. Talmi and D. T. Bostick, *Anal. Chem.* **47**, 2145 (1975).

50. Y. Talmi and A. W. Andren, *Anal. Chem.* **46**, 2122 (1974).

51. R. E. Kaiser, *Anal. Chem.* **45**, 965 (1973).

52. P. C. Uden, Specific Element Detection in Chromatography by Plasma Emission Spectroscopy, in *Environmental Speciation and Monitoring Needs for Trace Metal-Containing Substances from Energy-Related Processes*, F. E. Brinckman and R. H. Fish, Eds., NBS Special Publication 618, U.S. Government Printing Office, Washington, D.C., 1981.

53. J. W. Carnahan, K. J. Mulligan, and J. A. Caruso, *Anal. Chim. Acta* **130**, 227 (1981).

54. M. Morita, T. Uehiro, and A. Fuwa, *Anal. Chem.* **52**, 349 (1980).

55. D. M. Fraley, D. Yayes, and S. E. Manahan, *Anal. Chem.* **51**, 2225 (1979).

56. C. H. Gast, J. C. Kraah, H. Poppe, and F. J. M. J. Maessen, *J. Chromatogr.* **185**, 549 (1979).

57. G. E. Parris and F. E. Brinckman, *J. Org. Chem.* **40**, 3801 (1975).

58. G. E. Parris and F. E. Brinckman, *Environ. Sci. Technol.* **10**, 1128 (1976).

59. R. E. Mesmer and C. F. Baes, Jr., The Hydrolysis of Cations, Report No. ORNL-NSF-EATC-3, Oak Ridge National Laboratory, Oak Ridge, TN, 1974.

60. J. C. Chambers and B. E. McClellan, *Anal. Chem.* **48**, 2061 (1976).

61. A. D. Little, Co., Literature Study of Selected Potential Environmental Contaminants: Antimony and Its Compounds, U.S. Environmental Protection Agency, Washington, D.C. 1976.

62. Y. Talmi and V. E. Norvell, *Anal. Chem.* **47**, 1510 (1975).

63. C. C. Foreback, Some Studies on the Detection and Discrimination of Mercury, Arsenic, and Antimony in Gas Discharges, Ph.D. thesis, University of South Florida, 1973.

64. F. Challenger, *Chem. Rev.* **36**, 315 (1945).

65. National Research Council Report, Arsenic, National Academy of Sciences, Washington, D.C., 1977.

66. E. A. Woolson, Ed., "Arsenical Pesticides," ACS Symposium Series 7, American Chemical Society, Washington, D.C., 1975.

67. G. Stratton and H. C. Whitehead, *J. Am. Water Works Assoc.* **54**, 861 (1962).

68. S. A. Peoples, J. Lakso, and T. Lais, *Proc. West. Pharmacol. Soc.* **14**, 178 (1971).

69. R. S. Braman and C. C. Foreback, *Science* **182**, 1247 (1973).

70. D. L. Johnson and R. S. Braman, *Deep Sea Res.* **22**, 503 (1975).

71. W. R. Penrose, H. B. S. Conacher, R. Black, J. C. Merauger, W. Miles, H. M. Cunningham, and W. R. Squires, *Environ. Health Perspectives* **19**, 53 (1977).

72. J. S. Edmonds, K. A. Francesconi, and J. R. Cannon, *Tetrahedron Lett.* **18**, 1543 (1977).

73. E. A. Crecelius, *Environ. Health Perspectives* **19**, 147 (1977).

74. R. Kunkel and S. E. Manahan, *Anal. Chem.* **45**, 1465 (1973).

75. D. R. Jones IV and S. E. Manahan, *Anal. Chem.* **49**, 10 (1977).

76. R. J. Stolzberg and D. Rosin, *Anal. Chem.* **49**, 226 (1977).

77. R. G. Smith, Jr., *Anal. Chem.* **48**, 74 (1976).

78. Y. K. Chau and K. Lum-Shue-Chau, *Water Res.* **8**, 383 (1974).

79. Y. K. Chau and P. T. S. Wong, Complexation of Metals in Natural Waters, in *Workshop on Toxicity to Biota of Metal Forms in Natural Water*, Great Lakes Advisory Board, 1976.

80. P. G. C. Campbell, M. Bisson, R. Gagne, and A. Tessler, *Anal. Chem.* **49**, 2358 (1977).

81. *Methods for Chemical Analysis of Waters and Wastes*, Environmental Protection Agency, Cincinnati, OH, 1974.

82. M. S. Shuman and G. P. Woodward, Jr., *Anal. Chem.* **45**, 2032 (1973).

83. J. Buffle, F. L. Greter, and W. Haerdi, *Anal. Chem.* **49**, 216 (1977).

84. R. L. Wershaw, D. J. Pinckney, and S. E. Booker, *J. Res. U.S. Geol. Surv.* **3**, 123 (1975).

85. M. Schnitzer, M. I. Ortiz de Serra, and K. Ivarson, *Soil Sci. Soc. Am. Proc.* **37**, 229 (1973).

86. R. L. Wershaw and D. J. Pinckney, *Science* **199**, 906 (1978).

87. M. Schnitzer and S. U. Khan, *Humic Substances in the Environment*, Marcel Dekker, New York, 1972.

88. M. M. Kononova, *Soil Organic Matter*, Pergamon, Oxford, 1966.

89. K. Nishikawa and K. Tahata, *Bull. Tokai Reg. Fish. Res. Lab.* **58**, 233 (1969).

90. T. L. Shaw and V. M. Brown, *Water Res.* **8**, 377 (1974).

91. J. B. Sprague, *Nature* **220**, 1345 (1968).

92. P. Zitko, W. V. Carson and W. G. Carson, *Bull. Environ. Contam. Toxicol.* **10**, 265 (1973).

93. V. M. Brown, T. L. Shaw, and D. G. Sharben, *Water Res.* **8**, 797 (1974).

94. D. P. Chynoweth, J. A. Black, and K. H. Mancy, Effects of Organic Pollutants on Copper Toxicity to Fish, in *Workshop on Toxicity to Biota of Metal Forms in Natural Waters*, Great Lakes Advisory Board, 1976.

95. R. W. Andrew, Toxicity Relationships to Copper Forms in Natural Waters, in *Workshop on Toxicity to Biota of Metal Forms in Natural Waters*, Great Lakes Research Advisory Board, 1976.

96. W. J. Blaedel and D. E. Dinwiddie, *Anal. Chem.* **47**, 1070 (1975).

97. J. M. Wood, *La Recherche* **7**, 711 (1976).

98. P. T. S. Wong, Y. K. Chau, and P. L. Luxon, *Nature* **253**, 263 (1975).

99. R. M. Harrison and R. Perry, *Atmos. Environ.* **11**, 847 (1977).

100. E. Rohbock, H. W. Georgh, and J. Muller, *Atmos. Environ.* **14**, 89 (1980).

101. R. M. Harrison, R. Perry, and D. H. Slater, *Atmos. Environ.* **8**, 1187 (1974).

102. K. W. Olson and R. K. Skogerboe, *Environ. Sci. Technol.* **9**, 227 (1975).

103. P. D. E. Biggins and R. M. Harrison, *Environ. Sci. Technol.* **13**, 336 (1980).

104. W. R. Boggess, Ed., Lead in the Environment, Report No. NSF/RA-770214, Natural Science Foundation, Washington, D.C., 1977.

105. T. G. Movering, Ed., Lead in the Environment, U.S. Geological Survey professional paper 957, U.S. Government Printing Office, Washington, D.C., 1976.

106. National Research Council, *Lead in the Human Environment*, National Academy of Sciences, Washington, D.C., 1980.

107. L. J. Goldwater, *Mercury—A History of Quicksilver*, York Press, Baltimore, MD, 1972.

108. L. J. Goldwater, *Methyl Mercury in Fish*, Nordisk Hygienisk Tidskrift, Supplementum 4, Stockholm, 1976.

109. G. L. Baughman, J. A. Gordon, N. L. Wolf, and R. G. Zepp, Chemistry of Organamercurials in Aquatic Systems, Report No. EPA-660/3-73-012 U.S. Environmental Protection Agency, Corvallis, OR, 1973.

110. L. Fishbein, *Chromatogr. Rev.* **13**, 133 (1970).

111. R. C. Heaton and H. A. Laitinen, *Anal. Chem.* **46**, 547 (1974).

112. D. L. Rabenstein, C. A. Evans, M. C. Tourangeau, and M. T. Fairhurst, *Anal. Chem.* **47**, 338 (1975).

113. National Research Council, *An Assessment of Mercury in the Environment*, National Academy of Sciences, Washington, D.C., 1978.

114. G. Westoo, *Acta Chem. Scand.* **20**, 2131 (1966).

115. G. Westoo, *Acta Chem. Scand.* **22**, 2277 (1968).

116. K. Matsunaga, T. Ishida, and T. Odo, *Anal. Chem.* **48**, 1421 (1976).

117. C. J. Cappon and J. C. Smith, *Anal. Chem.* **49**, 365 (1977).

118. J. Toffaletti and J. Savory, *Anal. Chem.* **47**, 2091 (1975).

119. A. M. Kiemeneij and J. G. Kloosterboer, *Anal. Chem.* **48**, 575 (1976).

120. T. C. Rains and O. Menis, *J. Assoc. Off. Anal. Chem.* **55**, 1339 (1972).

121. A. A. El-Awady, R. B. Miller, and M. J. Carter, *Anal. Chem.* **48**, 110 (1976).

122. R. J. Baltisberser and C. L. Knudson, *Anal. Chem.* **47**, 1402 (1975).

123. R. O. Arah and B. McDuffie, *Anal. Chem.* **48**, 195 (1976).

124. J. J. Bisogni and A. W. Lawerence, *Environ. Sci. Technol.* **8**, 850 (1974).

125. P. Jones and G. Nickless, *J. Chromatogr.* **76**, 285 (1973).

126. W. Peters, *Ber. Dtsch. Chem Ges.* **38**, 2567 (1905).

127. P. Mushak, F. E. Tibbetts III, P. Zarnegar, and G. B. Fisher, *J. Chromatogr.* **87**, 215 (1973).

128. P. Zarnegar and P. Mushak, *Anal. Chem. Acta* **69**, 389 (1974).

129. P. Jones and G. Nickless, *J. Chromatogr.* **89**, 207 (1974).

130. S. J. Long, D. R. Scott, and R. J. Thompson, *Anal. Chem.* **45**, 2227 (1973).

131. D. L. Johnson and R. S. Braman, *Environ. Sci. Technol.* **8**, 1003 (1974).

132. R. S. Braman and D. L. Johnson, Analytical Studies on the Speciation of Ambient Levels of Mercury in Air, Proceedings, International Symposium on Environment and Health, Paris, 1974.

133. R. D. Rogers, Methylation of Mercury in a Terrestial Environment, Abstract C-218, International Conference on Heavy Metals in the Environment, Toronto, Canada, 1975.

134. P. E. Trujillo and E. E. Campbell, *Anal. Chem.* **47**, 1629 (1975).

135. L. L. Smith, Jr., D. O. Oseid, and L. E. Olson, *Environ. Sci. Technol.* **10**, 565 (1976).

136. J. A. Turner, R. H. Abel, and R. A. Osteryoung, *Anal. Chem.* **47**, 1343 (1975).

137. J. J. Renard, G. Kubes, and H. I. Bolker, *Anal. Chem.* **47**, 1347 (1975).

138. S. S. Brody and J. E. Chaney, *J. Gas Chromatogr.* **2**, 42 (1966).

139. D. G. Greer and T. J. Bydalek, *Environ. Sci. Technol.* **7**, 153 (1973).

140. R. K. Stevens, J. D. Mulik, A. E. O'Keefe, and K. J. Krost, *Anal. Chem.* **43**, 829 (1971).

141. D. F. S. Natusch, J. R. Sewell, and R. L. Tanner, *Anal. Chem.* **46**, 410 (1974).

142. B. J. Slatt, D. F. S. Natusch, J. M. Prospero, and D. L. Savoie, *Atmos. Environ.* **12**, 981 (1978).

143. S. O. Farwell, S. J. Gluck, W. L. Bamesberger, T. M. Shutte, and D. F. Adams, *Anal. Chem.* **51**, 609 (1979).

144. D. F. Adams, S. O. Farwell, E. Robinson, M. R. Pack, and W. L. Bamesberger, *Environ. Sci. Technol.* **15**, 1493 (1981).

145. D. F. Adams, S. O. Farwell, M. R. Pach, and E. Robinson, *J. Air Pollut. Control Assoc.* **31**, 1083 (1981).

146. H. R. Goering, E. E. Cary, L. H. P. Jones, and W. H. Allaway, *Soil Sci. Soc. Am. Proc.* **32**, 35 (1968).

147. G. Gissel-Nielsen and B. Bisbjerg, *Plant Soil* **32**, 382 (1970).

148. C. S. Evans, C. J. Asher, and C. M. Johnson, *Austrian J. Biol. Sci.* **21**, 13 (1968).

149. R. W. Fleming and M. Alexander, *Appl. Microbiol.* **24**, 424 (1972).

150. P. B. Dransfield and F. Challenger, *J. Chem. Soc.* **2**, 1158 (1975).

151. K. P. McConnell and O. W. Portman, *J. Biol. Chem.* **195,** 277 (1952).

152. K. P. McConnell and O. W. Portman, *Proc. Soc. Exp. Biol. Med.* **79,** 230 (1952).

153. J. L. Byard, *Arch. Biochem. Biophys.* **130,** 556 (1969).

154. I. S. Palmer, R. P. Gunsalus, A. W. Halverson, and O. E. Olson, *Biochem. Biophys. Acta.* **177,** 336 (1969).

155. National Research Council Report, *Selenium,* National Academy of Sciences, Washington, D.C., 1977.

156. S. F. Trelease, A. A. Disomme, and A. I. Jacobs, *Science* **132,** 618 (1960).

157. P. J. Peterson and G. W. Butler, *Austral. J. Biol. Sci.* **15,** 126 (1962).

158. B. G. Lewis, C. M. Johnson, and T. C. Broyer, *Biochem. Biophys. Acta* **237,** 603 (1971).

159. H. E. Ganther, *Biochemistry* (USSR) **7,** 2898 (1968).

160. H. E. Ganther, *Biochemistry* (USSR) **10,** 4089 (1971).

161. P. D. Goulden and P. Brooksbank, *Anal. Chem.* **46,** 1431 (1974).

162. E. L. Henn, *Anal. Chem.* **47,** 428 (1975).

163. G. Nelson, Analytical Methods or Aqueous Selenite Ion In the Environment, M.S. thesis, University of South Florida, Tampa, FL, 1976.

164. Y. K. Chau, P. T. S. Wong, and P. D. Goulden, *Anal. Chem.* **47,** 2279 (1975).

165. W. A. Aue and H. H. Hill, Jr., *Anal. Chem.* **45,** 729 (1973).

CAPILLARY GAS CHROMATOGRAPHY IN THE ANALYSIS OF ENVIRONMENT

MILOS NOVOTNY

Department of Chemistry
Indiana University
Bloomington, Indiana 47405

1. INTRODUCTION

Applications of gas–liquid chromatography to environmental problems have been numerous. Its use has been particularly intensified during the last 15 years. This situation has been primarily caused by the development of new highly sensitive and selective detectors as well as the advent of combined gas chromatography/mass spectrometry (GC/MS).

In terms of sensitivity, which is of a major concern to most, if not all, measurements of organic pollutants in our ecosystem, the capabilities of the electron capture detector or a mass spectrometer monitoring a selected ion are unparalleled. The necessity of measuring trace amounts of various pollutants on a routine basis has had a major part in forcing many research and control laboratories to accept these powerful detection methods in spite of the vast number of experimental problems involved. Although many element-sensitive gas chromatographic detectors are not quite at the sensitivity level of the electron capture detector, recent improvements in their analytical performance will also have a notable impact on pollution studies. Thus many pollutants can now be measured down to parts-per-billion and parts-per-trillion levels with reasonable accuracy and precision while using these specialized detectors. For example, today's pesticide residue laboratories would be hard to imagine without the existence of the electron-capture detector, flame photometric or thermionic detectors, or GC/MS techniques.

Although the necessity of sensitive and selective detection in air and water pollution studies has been recognized for some time, considerably less attention has been paid to the role of the chromatographic column until recently. This chapter intends to emphasize the crucial importance of the column in numerous environmental applications where its high performance makes the task of identification and quantitation considerably easier. Using capillary columns becomes mandatory in the analysis of complex environmental media. Indeed, a number of reports published during the last several years endorse this direction clearly.

In contrast to an almost immediate response to the development of selective detectors, there has been much reluctance to use capillary columns in environmental chemistry. Yet it is widely recognized that organic pollutants are most commonly encountered in complex matrices of other organic materials. The task of separation is further complicated by the numerous chemical reactions taking place in the ecological systems, for example, photodecompositions, biological degradations, and metabolic alterations of pollutants. In addition, the naturally occurring sample constituents are typically present in quantities far greater than those of measured pollutants. Thus an effective separation of the compounds of interest from interfering molecular species becomes the best remedy.

Although the use of a selective detector often alleviates the problem of separation (indeed, many selective detectors were developed for exactly this reason), the best possible separation should still be aimed for. First, in many *identification* studies through spectral methods, it is virtually mandatory to "look at only one compound at a time." Second, the performance of a selective detector to measure solute concentration is sometimes affected by the presence of an interfering compound even when such a compound is not detectable itself. Although the necessary extent of separation and possible compromises leading to method simplification must be judged on individual basis, it is logical to expect that a wider use of high-resolution GC shall result in fewer steps prior to the end determinations. Thus, in view of the available techniques, the use of selective detectors should now be looked upon as a complementary rather than competitive approach to the analysis of environmental mixtures.

High-resolution (capillary) GC has now rapidly evolved with the emphasis on both improved resolution and trace analysis capabilities. The majority of the earlier studies centered around the preparation and special uses of glass capillary columns. With most problems of suitable column technology solved at present, numerous efforts continue to couple such columns with effective concentration and sampling methods at one end and improved detection and ancillary techniques at the other end. Emerging automation and data treatment options will be adding further strengths to the method in due time.

2. OBJECTIVES OF CAPILLARY GC

Individual chemicals or mixtures of substances are released into our environment that originate from various industries, agricultural uses of chemicals, combustion processes, waste products of life, and so on. Many of these are directly hazardous to human life by interfering with man's normal metabolic processes or are indirectly hazardous by endangering various animal species and thus changing our ecosystem. Air and water pollution matters provoke particular concern in densely populated areas.

Environmental analytical chemistry is challenged with ever increasing demands for more efficient, reliable, sensitive, and rapid methods for monitoring pollutants. In the field of organic pollutant analysis, the recently started exploitation of the potential of high-resolution GC will soon be a well-documented case in point.

Why is capillary GC so important? From the numerous separation tools now available to a chemist, the capillary column is the most powerful one. Although restricted to relatively volatile substances, the current capillary columns can resolve up to several hundred components from a single sample. Through the combination with suitable detection devices or ancillary tools, virtually all resolved mixture components can be identified and quantitated. The range of volatility (extending up to molecular weights of around 800 daltons in some cases) covers the majority of separation problems in air pollution analysis and very many problems in water pollution. At this stage, high-performance liquid chromatography (HPLC), otherwise the best alternative for less volatile substances, has not been developed to the point where it is as valuable as high-resolution GC. This assessment is primarily made in respect to very complex mixtures of previously unknown composition.

A great many environmental analytical problems are associated with complex mixtures. Many important practical examples can support this statement. Just to mention a few, automobile exhaust gases are known to contain hundreds of compounds, as are the mixtures of polycyclic aromatic hydrocarbons adsorbed on airborne particulates. One of the most dangerous groups of industrial pollutants, polychlorinated biphenyls, is already being manufactured in the form of complex mixtures. In all three cases, a high degree of component resolution is required.

Efficient chromatography may be desirable even in the case of a limited number of pollutants. Here, capillary GC may eliminate tedious and time-consuming fractionation and cleanup procedures otherwise needed to remove the ballast material. Not only is the methodology simplified, but, importantly, compound losses due to cleanup are generally minimized.

A primary pollutant that has been emitted to air or a water system can usually undergo various chemical reactions affected by light, biological systems, the presence of naturally occurring substances, and so on. Such transformations and biodegradations lead to additional substances to be monitored. This is yet another reason to use extensively high-resolution chromatographic methods. The answers to How much? and In what form?

will be demanded even more frequently in the future in the fields of pollution control and environmental toxicology.

Perception of the merits of capillary GC in trace analysis has drastically changed over the years. The long-maintained general view that the capillary column is unsuitable for trace determinations is no longer justified. Advances in concentration and direct sampling techniques that were made during the last decade permit more effective utilization of the total samples. In addition, capillary columns are favored over the packed columns because they provide sharper output concentration profiles (chromatographic peaks). Thus both mass-flow and concentration-sensitive detectors will clearly benefit from this column type as far as their sensitivities are concerned. Whenever the detector background signals are affected by the column bleed, capillary columns are once again preferred.

Column inertness is an additional important aspect of trace analysis. Inertness of glass and fused silica columns toward the chromatographed solutes will be emphasized in this chapter. In this respect, the newer capillary column types are clearly superior to the conventional support-packed columns in most trace analysis problems.

Sample "fingerprinting," that is, linking a particular sample to its origin, will clearly be among the most important uses of capillary GC. The best techniques of this kind are those that provide as much specific data on a given sample as possible, preferably through a single analytical method. Both spectroscopic and separation methods were widely applied to fingerprinting in the past, yet none matches the extraordinary performance of capillary GC. Some applications of capillary chromatography, for example, in identifying oil spills, types of combustion processes, and profiles of pollutants, have already been demonstrated.

3. METHODOLOGICAL ASPECTS

3.1. Glass and Fused Silica Capillary Columns

Since the invention of the capillary column in the late 1950s (1), a majority of columns were first manufactured from metal tubes. At a later date, so-called support-coated open tubular columns (2, 3) added certain features desirable for some applications. Glass, an otherwise superior material for making capillary columns, was long neglected in spite of the initial success of Desty and co-workers in the earlier days of gas chromatography (4, 5). Following the first successful approaches to the preparation of glass capillary columns with stable films of polar stationary phases in the late 1960s (6–9), interest spread very rapidly to numerous laboratories. The advantages of the relative inertness of glass combined with high column efficiencies not available with other column types were clearly demonstrated in the cited publications.

Since the complex mixtures originating from tobacco smoke were successfully analyzed with glass capillary columns (7, 10), there could be little doubt even then that similar techniques should be successful in air pollution studies. As various new sample concentration approaches evolved with time, glass capillary columns started to assume an important role also in water pollution analysis. A series of publications by Grob and co-workers (11–13) from this area became illustrative of the capabilities of these new techniques.

Mastering the technology of glass capillary columns, an exceedingly difficult task for a beginner, was initially a very crucial step for those laboratories that had realized the potential for dramatically improved separations and decided to embark on new developments. As more articles about new developments in column technology were being published during 1970s, the number of enthusiastic followers has grown rapidly. A successful transition from packed to capillary columns, which has now become a reality for numerous laboratories worldwide, has been a direct result of: (1) improved understanding of the behavior of organic liquids on glass surface; (2) highly reproducible technology resulting from both systematic and empirical surface studies; (3) large-scale manufacture of suitable commercial instruments and high-quality capillary columns themselves; and (4) various forms of professional education in this "technique-oriented" approach. In 1979, the development of fused silica columns by Dandeneau and Zerenner (14) has further removed the "psychological barrier" for those who were uncomfortable with relatively fragile glass capillary columns. Although capillary GC has become a very common analytical technique during the last several years, it should be pointed out that a long induction period preceded today's wide utilization. The most important developments of that period will now be briefly reviewed.

Capillary tubes of various lengths and diameters can easily be made from ordinary glass tubes by means of a glass-drawing machine, originally described by Desty et al. (15) and now commercially available from more than one source. Desty's outstanding success with hydrocarbon mixtures (5) led others to consider glass, but the problems of surface wettability proved discouraging. The initial problems that had to be overcome all dealt with the fact that a great majority of organic liquids exhibit large contact angle on the high-energy glass surface and thus wet it poorly. Although some earlier studies (16, 17) had already called attention to this problem, it was not a commonly accepted fact until much later.

As it has long been known that a geometrical modification of surfaces (e.g., roughening) increases their wettability, the earlier approaches to inducing a homogeneous liquid film inside glass capillary columns were based on this phenomenon. Thus Grob (6, 7) succeeded in the preparation of efficient polar columns through carbonization of glass surface, whereas Novotny and Tesarik (8, 9) utilized gaseous corrosive agents to "etch" the inner wall of glass capillaries. The actual appearance of surface-corroded columns as well as the physical spreading of liquid films was later revealed through electron scanning microscopy (18). An extension of the corrosion techniques (8, 9) is the so-called whisker growth on the surface (19, 20); this technology is frequently used for the preparation of wide-bore (0.5–0.7 mm i.d.) glass columns. Another, more recent surface roughening method is the precipitation of barium carbonate on the surface, as suggested by Grob (21).

Somewhat different approaches to affecting wettability of glass surfaces involve chemical modifications of surface structures. Glass, as many siliceous materials, contains silanol groups on the surface. These groups can be modified through a variety of organic reactions, of which silylation has found by far the most extensive utilization. The attached organic layer is capable of altering significantly the basic surface characteristics of glass as reflected in the value of critical surface tension (22). This approach has been used to introduce onto the surface organic moieties that interact with certain structural features of the stationary liquid that is later coated on such a modified surface. Thus a homogeneous spreading of the liquid on the basis of structural compatibility is assured (23). The corresponding model presented in Figure 1 is generally applicable to phases and surfaces of different polarity.

Wettability problems in the preparation of glass capillary columns can now be largely solved by the aforementioned surface treatments or their combination. More recent research activities concerning the capillary sur-

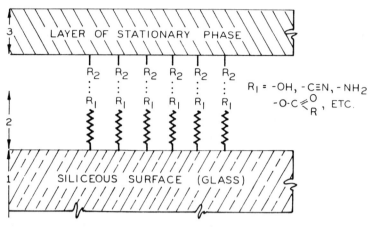

Figure 1. A model of chemically compatible surface structures. 1 = part of the glass silica framework; 2 = a permanently attached monolayer possessing a functional group R_1; 3 = a layer of stationary phase with a group R_2 that is chemically "compatible" with R_1.

face characteristics have concentrated on the removal of the so-called residual activity. Even though glass is known to be a considerably more inert column material than the previously used metals, there are many instances where improvements are needed. As a result of this residual activity, certain labile compounds tend to produce asymmetrical peaks, and when injected onto the column in very small quantities, they may "disappear" altogether. Irreversible adsorption of these molecules on the column surfaces is responsible for this phenomenon. Sensitive solutes may be further degraded on the surface containing catalytic sites.

The overall result of inadequate surface inertness is the loss of chromatographic resolution and poor quantitation. It is not uncommon, in such a case, for the trace analysis of certain compounds to be limited less by the detector sensitivity than the capillary column itself. In particular, these problems become visible when polar solutes are separated on nonpolar columns. Tailing of solutes, such as primary amines, pyridines, and alcohols, is experienced with insufficiently deactivated capillary columns.

Fortunately, various developments of the recent years have significantly reduced these column residual activity problems. Although much remains to be learned about the specifics of column deactivation processes, the main problems can be traced to (1) concentration of various glass ingredients in the surface and (2) reactivity of surface hydrated structures (e.g., silanol groups).

The importance of glass composition for good chromatographic results had been noticed quite early. Thus Filbert and Hair (24) considered removal of calcium ions from glass beads essential for improved chromatography, while Franken and Trijbels (25) suspected undesirable effects due to boron. Grob and Grob (26) called attention to the role of the pH of glass and devised the first testing procedure to assess the neutrality of column surface. The work of Lee et al. (27, 28) has provided a valuable insight into the surface deactivation processes through the detection of surface elements by Auger spectroscopy and their quantitation under different conditions and treatments. In order to remove these elements, acid leaching techniques have now been widely employed. These ingredients are practically absent on the inner surface of fused silica columns, and this is believed to be the primary reason for good inertness of such columns.

In order to reduce the undesirable surface interactions with a variety of solutes, numerous other approaches to surface deactivation have been proposed. The observations by Aue et al. (29) that siliceous surfaces retain a very small layer of a coated polymer after heat treatment and exhaustive solvent extraction have also been explored in capillary GC (30, 31). Such minute polymer layers spreading over the surface may "mask" undesirable interactions and prevent the solute molecules from entering the surface structures underneath. Numerous modifications of this approach now appear in the chromatographic literature, achieving successful deactivation for various analytical separations.

It has been observed for some time that a partial success in surface deactivation can be achieved through addition of certain surfactants to the stationary phase. For example, long-chain ionic compounds (32), a phosphonium salt (33), and a boranate (25) were shown to act as "tail reducers." Temperature stability of these treatments vary in dependence on a given surfactant.

The previously mentioned reactivity of the surface silanol groups may also be detrimental to chromatography of certain solutes. Surface silylation contributes, in general, to reduced activity of glass through blocking these reactive silanol groups and shielding various chromatographed molecules from the column's active sites. Although silylation with a variety of silyl donors has been used in glass capillary columns for a long time (8, 23), it was found more recently (34, 35) that the reaction carried out at high temperatures is essential for best deactivation results. Today, the high-temperature silylation is used widely. In addition, the use of cyclic siloxanes and silazanes for chemical modification of both glass and fused silica capillaries (36) bears a close resemblance to the general silylation approach.

Yet another recent trend in the capillary column technology has been the preparation of chemically bonded stationary phases. The idea of attaching pre-polymerized siloxanes to the column wall has been explored by Madani et al. (37) and Blomberg and Wännman (38). The advantage of this approach is primarily in connection with newer sampling techniques (splitless and on-column sample injections), as these column substrates are essentially nonextractable with a variety of organic solvents. Thus the undesirable phenomenon of liquid-phase stripping following the solvent injection is practically avoided. A similar effect is expected with the recently reported column technology (39) where the nonextractable polymeric layers are prepared through cross-linking the mechanically deposited ordinary phases.

It should be added, for the sake of completeness, that an equally important part of the capillary column technology is the method for depositing a uniform layer of the stationary film on the column wall. However, a coating technique is successful only if preceded by an appropriate surface treatment. If this is not the case, the film easily breaks up into droplets and lenses immediately on coating or while changing the column temperature. Coating methods were studied quite extensively (40–44) in terms of physical phenomena as well as by empirical means. Both static and dynamic coating procedures are widely employed, and a preferential use of one over the other depends on a particular column technology problem.

The section of this chapter dealing with the column technology matters, as just discussed, is only a brief survey of the literature published on this subject. This is intentional, because an environmental chemist is primarily interested in the uses rather than the column preparation. For a more detailed treatment of this subject, additional information may be sought elsewhere (45, 46) by an interested reader.

Capillary GC has become a very mature technique. The methods for preparation of high-quality columns are now widely available. Successful commercialization of this area has undoubtedly brought a certain degree of standardization during the last several years.

Although unusual phase selectivities are occasionally needed for special separations, only a limited number of column types is generally sought in capillary GC. Unlike work with packed columns, where the stationary-phase selectivity is crucial, there is not such a need with the capillary columns of high resolving power. The point expressed long ago by Giddings (47) that ". . . in the case of many component mixtures, such as various petroleum fractions, changes in selectivity may do little more than scramble the already crowded chromatogram, with new overlaps replacing the old; the only certain means of improving resolution is through an increase in the number of plates" is still valid and fits well the present situation with complex environmental mixtures. As discussed in the following sections of this chapter, other techniques combined with capillary GC can significantly contribute to the ultimate solution of a given analytical problem.

The effect of capillary inner diameter on column efficiency is predictable: the columns with decreased inner diameter have substantially increased plate numbers, while efficiency is compromised for a larger sample capacity with wide-bore columns. Capillary columns that are commonly used today have inner diameters between 0.2 and 0.3 mm. Instrumental difficulties, for example, problems of sampling and detector dead volumes, are generally encountered with smaller diameters. Both potential and difficulties of the small-bore columns were clearly demonstrated (4) more than 20 years ago. Surprisingly little has been done in this area since. Theoretical optimization of small-bore capillary GC was reported by Guiochon (48). A more recent experimental study shows some encouraging results (49), as demonstrated by the fast analysis of hydrocarbon mixture shown in Figure 2. A full development of this area will undoubtedly require significant departures from the present state of instrumentation.

Wide-bore (0.5–0.7 mm i.d.) capillary columns are also of some interest to environmental chemistry. This type of column is frequently preferable in cases when the column capacity becomes a problem, as the maximum sample amount that can be introduced without loss of efficiency is a square function of the column radius. While nanogram quantities are common for chromatographic peaks on 0.25-mm-i.d. columns, wide-bore capillaries can tolerate up to micrograms. Such amounts may be needed where the separated peaks are trapped for further investigation by the methods such as IR, Fourier-transform NMR, and selective chemical reactions. Wide-bore columns with the stationary phase coated directly on the wall have now effectively replaced support-coated open tubular columns that

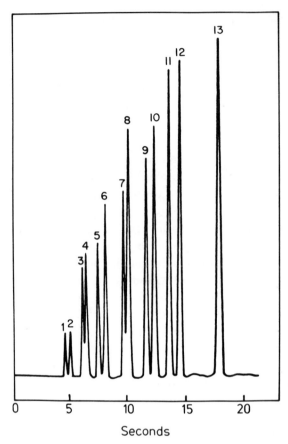

Figure 2. Fast capillary GC separation of a hydrocarbon mixture. 1 = *n*-pentane; 2 = 2-methylpentene-1; 3 = 4-methylpentene-1; 4 = 2,3-dimethylbutane; 5 = 2-methylpentene-1; 6 = *n*-hexane; 7 = methylcyclopentane; 8 = 2,4-dimethylpentane; 9 = benzene; 10 = cyclohexane; 11 = 2-methylhexane; 12 = 3-methylhexane; 13 = *n*-heptane. Column: 3 m × 30 μm i.d., fused silica capillary with a 0.06-μm film of a silicone gum. Reprinted from Schutjes et al. (49), with permission of the Institute of Chromatography, Bad Dürkheim, West Germany.

generally suffer from greater residual activity and the problems of irreversible adsorption on the solid support. The column technologies of wide-bore and the conventional (0.25-mm) columns may differ significantly. An extensive geometrical modification of the surface is generally required for the wide-bore columns. At this time wide-bore fused silica columns are not available.

An effective utilization of very efficient capillary columns requires appropriate instrumentation as well as the necessary know-how for work with such equipment. In addition, periodical tests of the column performance are needed to secure quantitative data. Today's technology readily yields columns with the plate-heights comparable to the column diameter as predicted from the theory. In addition to occasional measurements of the column efficiency, testing for residual activity is required. This is best accomplished through the use of various "molecular probes" designed to secure reliable quantitation for a particular analytical problem.

Various compounds that are difficult to chromatograph may serve as sensitive "molecular probes." If acid–base properties of the column wall are an important issue, a simple mixture of dimethylaniline and dimethylphenol, as suggested by Grob and Grob (26), may be a convenient sample. Some of the recommended tests may be as involved as in the chromatography of a recently advocated 13-component mixture (50), but the objectives of a given analysis should be met without undue hardship. An example of excellent chromatography is shown in Figure 3 where certain underivatized phenols (priority pollutants) have been eluted from a fused silica column coated with a methylsilicone gum (46) as symmetrical peaks. However, it must be emphasized that such peak symmetry should be maintained at the level of desired analyses. As demonstrated in Figure 4 (51), this may have serious consequences in trace analysis.

3.2. Sample Introduction

A prerequisite of any quantitative evaluation is that a representative sample is reproducibly introduced onto the capillary column. A direct introduction of nanogram samples is an obvious technological problem. To achieve the overall analytical objective, that is, to provide the detector with easily measurable solute quantities without overloading the capillary column, indirect sampling approaches are employed. However, the sample-splitting devices, which found a frequent use in the past for analyses of the major component in a mixture, are generally unsuitable in environmental applications. In the majority of such cases, sensitivity is a limiting factor and sample splitting becomes undesirable. In addition, splitting injectors are believed to be nonquantitative with respect to sample components of different boiling points (the so-called splitter discrimination).

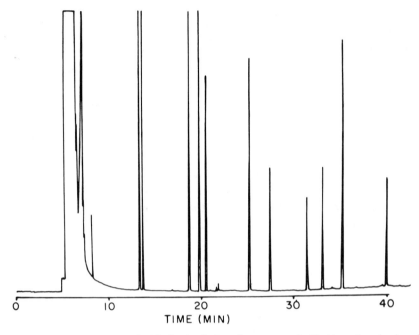

Figure 3. Chromatogram of underivatized phenolic compounds (2-chlorophenol, phenol, dimethylphenol, 2-nitrophenol, dichlorophenol, dichlorocresol, trichlorophenol, p-nitrophenol, dinitrophenol, dinitromethylphenol, and pentachlorophenol). Conditions: 15 m × 0.22 mm i.d., fused silica capillary coated with SE-30 methylsilicone gum; programmed 100–200°C at 8°C/min. Reprinted from Jennings (46) with permission of Academic Press.

Since most problems in environmental and biochemical analysis are related to trace components, many research groups have investigated techniques for direct sampling in capillary GC. The compounds of interest are usually encountered in dilute aqueous or gaseous media from which they must be preconcentrated on a solid adsorbent or in an organic solvent prior to the GC analysis.

Splitless injection techniques were developed for the analysis of trace components dissolved in a large excess of organic solvent. Initially Grob and Grob (52) reported a simple direct injection technique in which the less volatile trace components of a sample are "condensed" at the capillary column inlet, at relatively low temperatures, with minimal peak spreading; under these conditions, a large peak of a more volatile solvent passes through the column with little retention. In principle, this is a version of the cryogenic injection proposed earlier by Rushneck (53) for

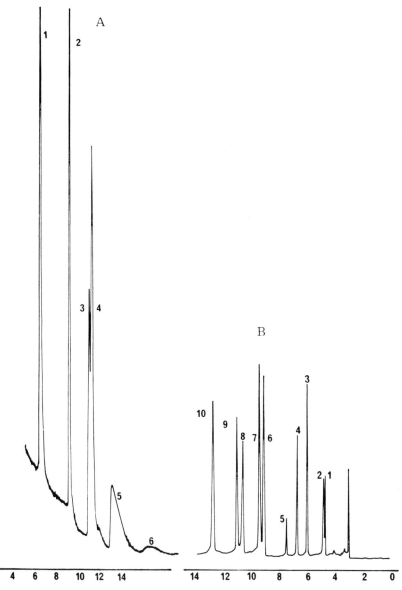

Figure 4. Chromatograms of a synthetic pesticide mixture, at 226°C, obtained with a 25.6 m × 0.22 mm i.d., glass capillary column coated with OV-101 silicone fluid, at (A) picogram level. Components: 1, γ-BHC; 2, heptachlorepoxide; 3, *pp'*-DDE; 4, dieldrin; 5, *pp'*-DDD; 6, *pp'*-DDT; (B) nanogram level. Components: 1, HCB; 2, γ-BHC; 3, heptachlor; 4, aldrin; 5, heptachlorepoxide; 6, *pp'*-DDE; 7, dieldrin; 8, *pp'*-DDD; 9, *op'*-DDT; 10, *pp'*-DDT. Reprinted from Franken and Rutten (51) with permission of Applied Science Publishers.

gaseous samples. The entire sample can be utilized following such procedures. Numerous technical aspects of this sampling are highly critical, as detailed in the later publications on the sampling mechanism (54, 55); these critical technical aspects are column temperature, boiling point of the used solvent, and sample dilution. A successful utilization of this injection technique is strongly dependent on the so-called solvent effect (54) under which a partial solvent condensation occurs just outside the hot injector, while the trace components are being effectively concentrated on the end part of the solvent zone. This phenomenon has now been utilized in a number of industrial and environmental applications.

More recently, alternative sampling methods have been developed (56–58) that facilitate direct injection of a liquid sample through a long, thin needle that extends inside the capillary column. These "on-column injection procedures" avoid some serious problems of the earlier sampling approaches: nonquantitative solute transfer from a heated syringe needle as well as the thermal and catalytic sample decomposition inside the vaporization chamber. Improved quantitation over the conventional injectors was particularly noted (58) with heavier sample components.

Another frequently encountered problem in the trace analysis by capillary GC is that capillary columns will typically tolerate only a small portion of available samples. For example, if a 1-μL aliquot of a 100-μL total sample is the maximum amount that a given column could tolerate, such a sample is not effectively utilized in a high-sensitivity measurement. The sampling procedures that are capable of overcoming this problem are the dropping-needle technique (59), the use of a precolumn (60), and "selective sampling" approaches (61). Such sampling devices are likely to find increasing utilization in future work.

It should be emphasized at this point that the sample introduction is only one of the many factors entering the admittedly difficult task of quantitation in complex environmental samples. Although some sampling approaches are deemed adequate today for the purpose of certain applications, further technological improvements are still desirable. Automation of sample introduction has proved to be beneficial in keeping good precision, in some cases well below 1%.

Naturally, the overall successful quantitative analysis is a multifaceted problem of sample choice, quantitative recovery during extractions and sample purification steps, choice of internal standards, signal recording technology, and so on. Most of these topics are beyond the scope of this chapter, as they are not deemed to be particularly characteristic of capillary GC. Thus only column and sample introduction aspects were reviewed in the preceding pages, although it should be noted that even those may sometimes be strongly application-oriented. In addition, some aspects of sample preconcentration will be mentioned in connection with the selected applications of capillary GC to air and water pollution problems.

3.3. Use of Different Detectors

Various selective detectors have been essential to many environmental applications of chromatography for a long time. Therefore, it is hardly surprising that they acquire special meaning in high-resolution chromatographic analysis. Selective detectors provide useful, complementary information to the complex chromatograms obtained with the universal flame ionization detector. In particular, certain element-specific detectors may provide qualitative indications that are essential even when the powerful GC/MS will ultimately be used. The information that a particular chromatographic peak is due to a substance containing sulfur, phosphorus, nitrogen, a halogen, or some other element becomes particularly useful during the task of structural elucidation.

Another important fact is that many selective detectors are more sensitive than the conventional flame ionization detector. Thus many important compounds that would be overlooked as peaks of negligible size or even undetected by a less sensitive detector frequently can be pointed out in a complex chromatographic profile. Many environmentally important substances and their metabolites are detectable by one or more members of the family of selective detectors.

A special role of selective detectors in conjunction with capillary columns has become apparent in profiling or fingerprinting samples of environmental interest. It has been shown that both sulfur- and nitrogen-sensitive detectors can provide important complementary information in identifying the sources of certain samples. An example of such fingerprinting application is demonstrated in Figure 5 where the same series of samples obtained from different automobile engine oils (62) was first recorded with the flame ionization detector and then with the thermionic (nitrogen-sensitive) detector.

Suitability of a detector for the capillary column work is determined

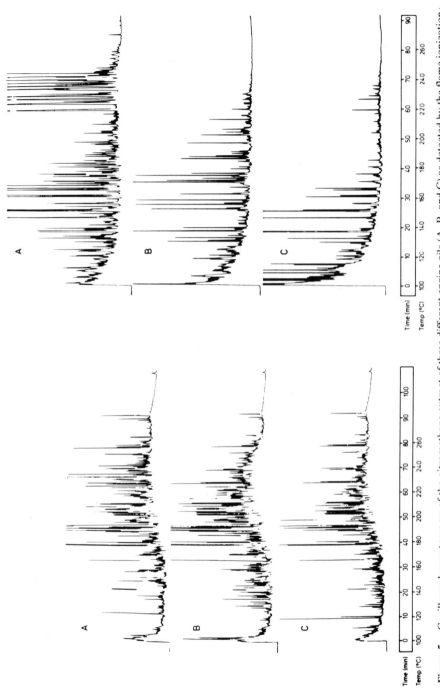

Figure 5. Capillary chromatograms of the nitromethane extract of three different engine oils (A, B, and C) as detected by the flame ionization detector (left) and a nitrogen-sensitive detector (right). Conditions: 22 m × 0.26 mm i.d., glass capillary column coated with SE-52 silicone polymer. Reprinted with permission from M. L. Lee et al., *Anal. Chem.* **47**, 540 (1975). Copyright 1975 American Chemical Society.

by the (1) small volume of the detection cell, (2) sensitivity compatible with the conditions of separation, and (3) fast detector response. Strictly taken, only few currently available detectors would qualify. With the exception of flame detection devices, dead volumes of interconnecting lines and the detector itself are the most serious problems; these can only be overcome or reduced by adding extra carrier gas at the column exit. Naturally, this approach leads to a certain sacrifice in the detection sensitivity of the concentration-sensitive detectors. The suitability of several detectors for capillary GC and environmental analysis will now be briefly discussed.

3.3.1. Thermionic Detector

The thermionic detector, often referred to as an "alkali flame ionization detector," has received considerable attention since its first description (63). Its utility was greatly reduced by insufficient technology and by design problems for a number of years. The detector bases its selectivity on the secondary ionization processes occurring in the flame in the presence of an alkali metal. However, these phenomena appear to be quite different for various designs and modes of operation and no straightforward explanations for the various mechanisms of detection currently exist. Various detector designs were observed to give enhanced responses to compounds containing phosphorus, halogens, arsenic, sulfur, tin, and some less common elements. Although many aspects of this type of detection were discussed (64, 65) in some detail, much remains to be explained. Future research on the mechanism of response may lead to improvements in both the sensitivity and selectivity of these promising element-selective devices. Environmental analytical chemistry could benefit greatly from such developments.

The most developed versions of the thermionic detectors have been those for selective analysis of halogenated, phosphorus-, and nitrogen-containing compounds. Although halogenated compounds are detectable in very small amounts with the electron capture detector, and the flame photometric detector provides an additional alternative for phosphorus compounds, the major importance of the thermionic detector seems to lie, at present, in the detection of nitrogen-containing compounds. Alternatively, an element to which the thermionic detector responds can often be incorporated into the molecules of interest by a suitable derivatization method provided that sufficient sample volatility is preserved.

The serious objections of the past against the thermionic detector were insufficient reliability and a limited signal stability. The detector versions incorporating the alkali source on the electrode loop or directly on the flame jet proved to be somewhat unreliable and certainly not adequate for day-to-day routine analytical use. As demonstrated by Kolb and Bischoff (66), a controlled external heating of the alkali source appears to be a good solution to the instability problems. The mechanism of detection has been quite satisfactorily explained by the interaction of cyano radicals with the excited rubidium atoms (66); the nature of response to phosphorus compounds with this detector is not known at present. Hartigan et al. (67) have further shown that selectivity of this detector is of the order of 10^3–10^4, sensitivity about 10^{-13} g/sec (approximately 50 times more sensitive than conventional flame ionization detectors), and the linearity range is comparable to that observed with the flame ionization detector. Because of its good long-term stability, this detector is quite suitable for routine analyses.

Since the thermionic detector is a type of flame detector, the rules that apply to its coupling with capillary columns are similar to those for the flame ionization detector, and no special modifications are needed. In combination with glass capillary columns, the nitrogen-sensitive detector was used to detect nitrogen-containing polynuclear aromatic compounds in the extracts of engine oils (62), airborne particulates (67), and other environmental samples (68). Nitrogen-containing polycyclic aromatic compounds, the so-called aza-arenes, are of environmental interest because of their carcinogenicity (69).

3.3.2. Flame Photometric Detector

The flame photometric detector is based on observing characteristic light emission subsequent to combustion of certain molecules in an energetic flame. Although it is also possible to apply this principle of detection to volatile metallic compounds (70), the initial work of Brody and Chaney (71) was primarily concerned with selective detection of sulfur and phosphorus compounds. In this latter fashion, it also retains its present role in environmental analysis. Whereas detection of phosphorus compounds is also possible with the thermionic detector, the detector's response to sulphur compounds at 394 nm (attributed to S_2 spectral species) is a unique capability. This is of great importance, since the presence and effects of sulfur compounds in petroleum products and coal fluids are well known.

Sulfur-containing polycyclics were also noted in airborne particulates (72, 73) and in carbon black used as a filler for car tires (74). In addition, some volatile sulfur compounds may strongly contribute to various odors and malodors in air pollution. The fingerprinting value of this detector has been shown by Adlard et al. (75) through monitoring crude oils and environmental samples associated with oil spills.

Unlike with the flame ionization detector, the hydrogen-rich flame is employed with this detector, and the column effluent is premixed with oxygen or air. In spite of this, the capillary column exit can be brought quite easily into the stream of combustion gases with not adverse effects on separation efficiency.

In fact, when dealing with trace amounts of sulfur compounds from complex organic sample matrices, maximum separating efficiency should be sought prior to this type of detection because of the phenomenon first observed by Rupprecht (76) and Perry and Carter (77). According to these authors, the presence of other organic compounds eluting simultaneously from the column with a sulfur compound under detection may result in a diminished response or quenching effect. Hence the need for efficient solute separation.

It should be noted that the response of the flame photometric detector to sulfur compounds is not linear, but it increases with the square of the sulfur content (77). With most detector designs, analyses can typically be performed at the low nanogram level.

3.3.3. Electron Capture Detector

Utility of the electron capture detector in environmental analysis hardly needs to be emphasized. The importance of this detector has also been strengthened more recently by overcoming the earlier drawbacks of the radioactive source instability and response nonlinearity. In addition, a proper understanding of the detection mechanisms (78, 79) has been a most welcome advance.

There are several reasons why coupling the electron capture detector with capillary columns is highly desirable. First, in the crude complex mixtures of environmental origin, many interfering compounds in addition to those under measurement are likely to occur. Unless laborious methods for sample "cleanup" are employed, the final measurement is frequently unreliable. On the other hand, if extensive sample purification is applied to assure specificity of a packed-column analysis, uncontrolled losses of trace compounds may occur. Obviously many interfering substances can readily be resolved from a sample of interest with capillary columns. Second, the great sensitivity of this detector is considerably enhanced when highly inert glass or fused silica columns are used. Minimum irreversible adsorption and sample losses result in easier detection and quantitation of electron-absorbing molecules. Also, capillary columns produce considerably less bleeding than the packed columns, thus reducing the detection problems under temperature programming. As a combined effect of these improvements, interesting environmental applications of the electron-capture detector at the subpicogram levels will become increasingly feasible.

Coupling capillary columns to an electron-capture detector is not entirely without problems. These arise primarily from certain constructional features of ionization detectors housing a radioactive source. Unlike with the flame detectors, relatively large detector cells are common, particularly with the older designs of the electron capture detector. Consequently, remixing and band broadening in the detector can seriously affect the separation ability of capillary columns. Although the use of a scavenger gas can reduce or eliminate this problem, it will negatively effect the detection response, depending on a particular detector design or voltage conditions (80).

A second approach to an optimum marriage of the electron capture detector, that is, to detector miniaturization, is also feasible (81, 82). An obvious benefit of this approach is a significant sensitivity enhancement due to the concentration-sensitive nature of this detector. In theory, this enhancement can be as much as one to two orders of magnitude. It has been shown (82) that a successful coupling can be achieved with a minimum loss of the column's theoretical plates.

Many excellent environmental applications of capillary columns in conjunction with the electron capture detector can now be found in the literature. Separations of chlorinated pesticides (83) and analyses of river water (84, 85) and airborne particulates (86) are representative examples. A significant rationale for the parallel use of the flame ionization and electron capture detectors for environmental samples is clearly demonstrated in Figure 6. The runs from a nonpolar fraction of the sewage water extract (82) are compared here to show the complementary nature of the two detector signals.

It is commonly believed that the potential of the electron capture de-

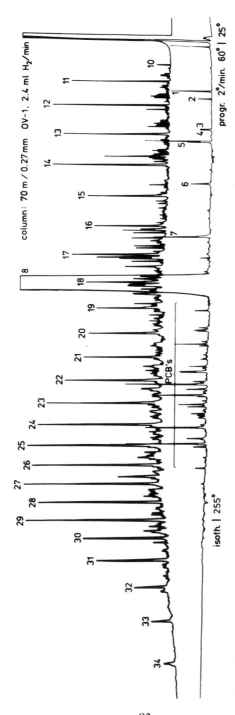

Figure 6. Analysis of the nonpolar fraction of extract from sewage, recorded simultaneously with flame ionization (upper trace) and electron capture (lower trace) detectors. Column: 70-m × 0.27-mm-i.d. glass capillary coated with OV-1 silicone phase. Reprinted from Grob (82) with permission of Pergamon Press.

83

tector has not yet been entirely explored. For example, recent developments in response sensitization (87, 88) seem to endorse such a belief. Further improvements are likely to come with advancing knowledge of the ion–molecule reactions.

3.3.4. Photoionization Detector

The photoionization detector is one of the oldest GC ionization detectors (89). Although different types of photoionization detectors were studied for a variety of reasons in several scientific laboratories, the practical aspects have not received enough attention until recent years. The sealed-source technology developed by Driscoll et al. (90, 91) has now made the photoionization detector more attractive as a reliable tool of trace analysis. Subject to the solution of seemingly solvable technological problems, this type of detector could well become one of the most important.

During the photoionization effect, a photon of sufficient energy (higher than the ionization potential of the irradiated molecular species) causes a release of electron: $M + h\nu \rightarrow M^+ + e^-$. If the molecules are encountered in a confined space (the actual adjunct "detector cell"), changes in conductivity can now be measured through the collection of electrons or, alternatively, ions. Unlike in the photoionization mass spectrometry, the fate of M^+ is of no particular concern in the photoionization detection.

Constructional features of the photoionization detector are quite simple (92), as exemplified in Figure 7: a UV lamp is attached to a small detector cavity with the two electrodes. Under stable conditions of the energy source, the relationship between the detector response and concentration can be remarkably linear, next only to the flame ionization detector. Depending on certain technological parameters of the detector cell, some photoionization detectors (93) were believed to be up to three orders of magnitude more sensitive than the flame ionization detector. Although such a sensitivity margin does not seem to be the case in the newer detectors with a UV sealed source (91), high sensitivities are still encountered with different organic molecules. Perhaps most importantly for the environmental applications, the photoionization detector provides very high sensitivity to some compounds for which the flame ionization detector performs poorly: halogenated compounds, vinyl chloride, tetraethyl lead, CS_2, H_2S, and so on. Its response is also "selectively" en-

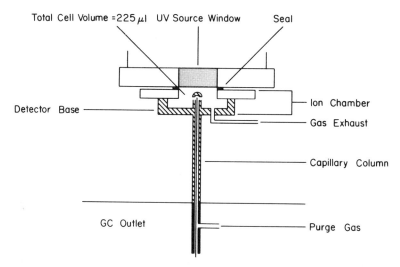

Figure 7. Schematic of a photoionization detector. Reproduced from Jaramillo and Driscoll (92) with permission of Hüthig Publishing Company.

hanced for aromatics, carbonyl compounds, and solutes containing heteroatoms. Using lamps of different energy (91) has some further implications in the response enhancement.

Some preliminary work on connecting the photoionization detector to a capillary column has already been reported (92). Since the detector shown in Figure 7 has a cell volume of 225 μL, reduction of its size is desirable. The chief reason, of course, is the necessity of avoiding dead volumes and the subsequent loss of efficiency. As shown in Figure 8, over 50% column efficiency has been lost with this detector cell at 1.6 mL/min flow rate while using a 0.25-mm-i.d. capillary column.

One of the interesting features of the photoionization detector is that it is a concentration-sensitive device. Thus a decrease in flow rate will increase the signal substantially for the same compound mass. Further miniaturization of this detector will be essential in order to have the benefits of high sensitivity and still preserve the column efficiency. Alternatively, a purge gas added at the column's end can overcome the dead volume problems, but only at a sacrifice of sensitivity. The situation here is quite similar to that of the previously discussed electron-capture detector, as well as the following discussion on UV spectroscopic detection.

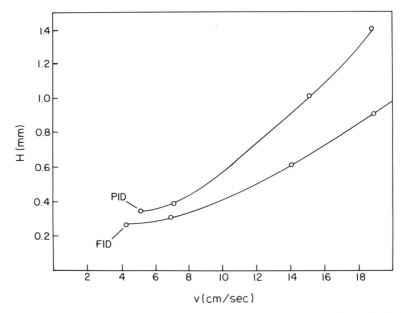

Figure 8. Van Deemter plots obtained with a photoionization (PID) and flame ionization (FID) detectors with the same capillary column. Reprinted from Jaramillo and Driscoll (92) with permission of Hüthig Publishing Company.

3.3.5. *Ultraviolet Absorbance Detector*

Various optical spectroscopic detectors comprise an important class of detection devices used in the contemporary chromatographic analysis. However, their utilization is almost exclusively confined to high-performance liquid chromatography (HPLC), although interesting possibilities of the gas-phase detection have been demonstrated throughout the years (94–98) for both the absorption and emission modes and wide spectral ranges. Selective monitors operating at an adjustable wavelength are potentially of some interest to environmental applications.

Although many technological improvements are still desirable, there are some indications (98, 99) that a wider utilization of such detectors in capillary GC may become feasible. The current technological problems concern mainly the design of miniaturized heated cells and some minor improvements of the optical arrangements.

Combination of a gas-phase UV detector with wide-bore (0.7-mm-i.d.) glass capillary columns was recently demonstrated (99). Figure 9 shows

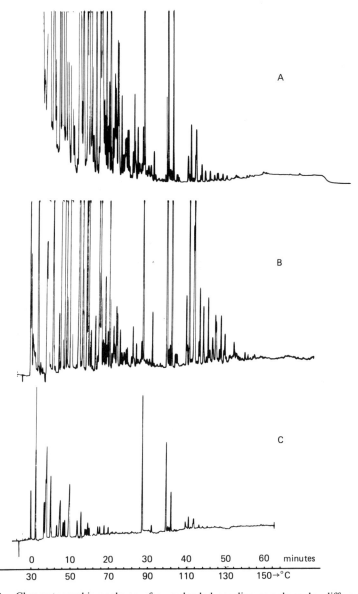

Figure 9. Chromatographic analyses of an unleaded gasoline sample under different detection conditions: (A) flame ionization detector; (B) the UV detector adjusted to 220 nm; and (C) the UV detector adjusted to 260 nm. Column: 32 m × 0.7 mm i.d., glass capillary column coated with a polypropylene glycol stationary phase. Reprinted with permission from M. Novotny et al. *Anal. Chem.* **52**, 736 (1980). Copyright 1980 American Chemical Society.

an application of this coupling to monitoring the aromatic compounds in a gasoline sample. Although every organic component within the given volatility range is detected by the flame ionization detector, olefins and aromatics are likely to be responsible for the peaks detected at 220 nm by the UV detector; at 260 nm, this detector should virtually "ignore" anything but the aromatic sample components. A 50-μL detector was used in this application; conventional 0.25-mm-i.d. capillary columns can also be used, but only with the purge gas to the heated cell and the correspondingly lowered sensitivity. However, future work on the instrumental design may overcome this difficulty.

3.4. Ancillary Techniques for Identification Purposes

Capillary gas–liquid chromatography is often (but not always) capable of a complete resolution of the individual mixture components. Our capabilities to positively identify such components are generally proportional to the degree of separation achieved. When considering the usual analytical methods that are available for structural elucidation, only a few of them will qualify because of the sample requirements. Although the situation with some spectral methods has drastically improved with a wider use of Fourier-transform techniques, trapping the capillary column effluents is not a reasonable alternative in a great majority of cases, even when technically feasible with either wide-bore wall-coated or support-coated columns.

Since the capillary chromatographic peaks are typically separated by seconds or less, on-line ancillary methods are mandatory. In particular, an enormous current interest in capillary GC/MS attests to the power of such on-line combination. Modifications of enrichment devices for the capillary column work and a later use of high-capacity pumping systems that can handle gas flows from capillary columns without a molecule separator have amply demonstrated that no sacrifice in the chromatographic performance of the combined method is needed. Engineering advances in various technical aspects of mass spectrometry have also been invaluable. Today, most instrument manufacturers consider an effective performance of their mass spectrometers with capillary columns essential.

The use of GC/MS in environmental laboratories has been increasing in spite of the equipment costs. Most instruments that are combined with capillary columns have been of the low-resolution type, although the current sensitivities of some double-focusing mass spectrometers are compatible with high-resolution columns. Acquisition of the exact masses, wherever appropriate, can be a significant benefit in identification work in spite of the enormous instrument costs. Today's modern mass spectrometers are equipped with powerful computers for data acquisition and reduction. In addition, extensive libraries of spectra obtained from environmentally important substances are common. Thus computers can also be used in matching the unknown acquired spectra with those of previously analyzed substances.

Clearly, acquisition of a "good mass spectrum" does not always lead to a positive identification. There are many instances when mass spectrometry alone fails to provide the necessary information. Optical spectroscopic methods, and particularly IR spectroscopy, can yield complementary data. However, it is more difficult to obtain spectra of a sufficient quality from nanogram sample quantities with such methods. Recent indications on Fourier-transform IR spectroscopy (100, 101) and optical imaging detectors (102) give some hope for progress. If further technological advances in this area take place, we may see a powerful array of identification techniques in the near future.

Many researchers have already recognized that GC/MS is not merely a combination of the two techniques, but rather a powerful tool in itself. This is best illustrated with many examples of distinguishing isomers: Although the mass spectrometer has difficulty in distinguishing isomeric compounds, these isomers may elute at different retention times. This point is particularly valid with capillary columns that can successfully resolve even very close isomers. Although various positional isomers, cis-trans pairs, diastereoisomers, and so on may yield practically identical mass spectra, their retention in capillary gas chromatography is frequently different.

Examples of the complementary information from mass spectra and retention data are encountered with polycyclic aromatic compounds. Their mass spectra may be nearly identical, yet differences in the chromatographic mobility are distinct. Thus, among the polynuclear aromatic hydrocarbons, most compounds with different molecular shapes can be readily distinguished; phenanthrene is eluted slightly before anthracene, benzo[j]fluoranthene before benzo[k]fluoranthene, and three isomers with molecular weight 252 are eluted in the following order: benzo[e]pyrene < benzo[a]pyrene < perylene. Likewise, differently methylated aromatic hydrocarbons exhibit significantly different retention on a glass capillary column (73). Distinguishing such compounds in various combustion products is very important because of their different toxicities (103).

Utility and any future extensive use of the relations between structure and retention will depend on (1) acquisition of a significant number of reference compounds for comparative purposes, (2) advancing predictions for retention, and, (3) high-precision measurements with capillary gas chromatography. Some progress has already been made. With reliable capillary column technology facilitating high-precision measurements, retention data on over 200 polycyclic aromatic compounds were obtained (104). Through the use of appropriate polycyclic aromatic standards, such measurements were shown to be better than ± 0.25 index units. As shown in Table 1, this approach is also applicable to nitrogen-containing aromatics (105), and, presumably, other classes of environmentally important substances (e.g., various isomeric pesticides, PCB's, and dioxins).

Other approaches to distinguishing various isomers could benefit from selective ionization techniques applied to mass spectrometry. For example, Lee and co-workers have recently demonstrated (106) that the mixed charge exchange–chemical ionization mass spectrometry can result in significantly different ratios of $(M + 1)/M$ ions within various polynuclear aromatic isomers. These ratios can further be correlated with ionization potentials that are theoretically predicted.

Even with the best available chromatographic efficiencies of several hundred thousand theoretical plates, certain components of complex mixtures still overlap. Additional means of simplification must be sought for capillary GC. The methods receiving increasing attention are (1) selective fractionation, (2) multiple-column chromatography, and (3) a wider use of selective GC detectors.

3.5. Evaluation of Chromatographic Profiles

Whereas identification of new or suspected compounds through the methods outlined in the preceding sections may be the primary goal of a given environmental study, there are numerous instances when the analysis of most, if not all, possible components of a chromatographic profile is highly desirable. Fingerprinting oil spills, monitoring efficiency of sewage treatment systems, and analyzing products of various combustion processes

TABLE 1. "Retention Indices" of Alkylated Pyridines, Quinolines, and Related Substances[a]

Compound	Average Index Value	Standard Deviation
A. Column: 50 m × 0.25 mm i.d., Glass Capillary Column Coated Statically with UCON 50-HB-2000 and Kalignost		
2-Methylpyridine	106.48	0.11
2-Methylpyrazine	111.04	0.16
2,6-Dimethylpyridine	113.75	0.17
3-Methylpyridine	115.49	0.24
4-Methylpyridine	116.26	0.20
2-Ethylpyridine	117.65	0.21
2,5-Dimethylpyridine	123.36	0.23
2,4-Dimethylpyridine	124.64	0.21
2,3-Dimethylpyridine	127.35	0.22
3-Ethylpyridine	130.17	0.25
4-Ethylpyridine	131.76	0.24
2,4,6-Trimethylpyridine	132.89	0.25
3,4-Dimethylpyridine	141.96	0.22
4-Vinylpyridine	142.77	0.23
2(3-Pentyl)pyridine	146.68	0.22
4-*t*-Butylpyridine	152.59	0.20
3-Ethyl-4-methylpyridine	155.36	0.21
2-Aminopyridine	192.62	0.06
2-Amino-6-methylpyridine	197.81	0.32
B. Column: 20 m × 0.25 mm i.d., Glass Capillary Column Coated Statically with UCON 50-HB-2000 and BTPPC		
8-Methylquinoline	207.25	0.10
2-Methylquinoline	207.88	0.11
2,8-Dimethylquinoline	211.86	0.09
8-Ethylquinoline	216.86	0.05
6-Methylquinoline	218.00	0.20
3-Methylquinoline	218.53	0.14
2-Isopropylquinoline	220.54	0.17
4-Methylquinoline	223.76	0.12
6,8-Dimethylquinoline	224.74	0.09
2,7-Dimethylquinoline	225.21	0.08
8-*n*-Propylquinoline	228.72	0.04
2,6,8-Trimethylquinoline	228.75	0.08
4,8-Dimethylquinoline	230.04	0.10

91

TABLE 1. (*Continued*)

Compound	Average Index Value	Standard Deviation
2-*n*-Propylquinoline	230.98	0.09
2,4-Dimethylquinoline	231.06	0.13
2,3-Dimethylquinoline	231.54	0.08
3-Ethylquinoline	232.62	0.14
6-Ethylquinoline	232.65	0.10
2,4,8-Trimethylquinoline	233.98	0.09
2-Ethyl-3-methylquinoline	238.34	0.16
2-Ethyl-3-methylquinoline	238.35	0.14
6,7-Dimethylquinoline	243.59	0.20
3,5,8-Trimethylquinoline	245.41	0.06
6-*n*-Propylquinoline	245.82	0.07
5,6-Dimethylquinoline	246.53	0.13
4-*n*-Propylquinoline	246.90	0.10
2,4,6-Trimethylquinoline	247.39	0.09
2,4,7-Trimethylquinoline	247.50	0.08
3,4-Dimethylquinoline	248.70	0.12
2,4-Diethylquinoline	249.09	0.05
2,4,6,8-Tetramethylquinoline	249.50	0.13
2,6,7-Trimethylquinoline	249.88	0.11
3-Methyl-4-ethylquinoline	253.73	0.11
3,7-Dimethylquinoline	254.36	0.12
2-Ethyl-3,5-dimethylquinoline	257.45	0.06
2-Ethyl-3,5,8-trimethylquinoline	260.21	0.04
2,3,4-Trimethylquinoline	260.61	0.08
3,4,7-Trimethylquinoline	264.91	0.04
3,6-Dimethyl-4-ethylquinoline	267.30	0.09
2,4,6,7-Tetramethylquinoline	270.29	0.08
3,4,5,8-Tetramethylquinoline	279.49	0.04
3,4,6,7-Tetramethylquinoline	287.64	0.12
2,3,4,6,7-Pentamethylquinoline	297.23	0.11

[a] Pyridine, quinoline, and acridine were assigned the index value of 100, 200 and 300, respectively. Data taken from Reference 105.

can be quoted as examples of such a situation. Often, quantitative rather than qualitative changes may be reflected.

Capillary GC and its related techniques lend themselves ideally to the fingerprinting analysis because of the wealth of analytical detailed information they may provide. Although so much information may seem confusing at first sight, it is an important task to reduce these "fingerprints" to some meaningful information. This can be accomplished either visually (if possible) or through the computer-aided evaluation of numerous chromatographic profiles.

The preceding considerations are not too remote from certain attempts in biomedical sciences where chromatographically recorded "metabolic profiles" are potentially useful in distinguishing various disease conditions (107–112). Here, again, high resolving power of capillary chromatography is necessary to resolve numerous metabolic products of the living organisms. Automation, reliable quantitative measurements, and the computer evaluation of very complex chromatographic data from numerous samples (109–113) are all essential.

A fully automated system has been described (113) for the analysis of volatile substances. It is capable of routine analysis of numerous samples per day, with reproducible recording of up to several hundred components in a chromatographic profile. Special-purpose instruments of similar nature could be designed for environmental and fingerprinting studies. The system's diagnostic power could be strengthened by the simultaneous operation of several selective detectors (113, 114) if necessary.

In order to deal with the enormous complexity of chromatographic profiles, a wider application of computer-aided evaluation methods will eventually become a necessity. Recent advances in the use of trainable pattern classifiers in chemistry and other scientific disciplines appear particularly suitable for the reduction and analysis of complex chromatographic data from many samples of a different type.

Applications of the pattern recognition computational methods to GC of less complex mixtures have been demonstrated (115) earlier. Such methods can be useful in (1) recognition of the common features in a chromatographic profile of a given sample and those previously analyzed and (2) identification of the actual molecular dissimilarities through the use of feature extraction programs. Both these advantages have been realized recently in the studies of molecular differences due to human diabetes (109, 110); the binary classifiers have shown up to 94% prediction accuracy (109) in the pattern distinction, while the alterations of important metabolites could be traced (110) following the feature extractions.

4.   SELECTED APPLICATIONS

No comprehensive account of the applications of capillary GC to pollution analysis is intended in this chapter. Several representative applications to air and water pollution studies are included to demonstrate the practical utility and advantages of this method. Some related analytical aspects will also be discussed briefly.

4.1.   Air Pollution

Many organic compounds are emitted daily to air in a number of ways such as escaping from industrial processes, burning of fuels, and operation of motor vehicles but also from various natural sources. Depending on a site of sampling, their amounts may range from parts-per-million down to parts-per-trillion levels. Some organic substances may further be adsorbed on airborne particulate matter.

Volatility of the analyzed samples is, of course, the first requisite for using GC. However, the use of the term "volatile pollutants" may be quite arbitrary. To avoid confusion within this chapter, we will reserve the meaning of "volatility" for the substances that are airborne (not associated with particles) and can be sampled directly from the polluted air and chromatographed in the gas phase without chemical derivatization.

Direct injection of samples of the polluted air into capillary columns is not feasible because (1) mixture components with relatively low vapor pressures are not easily detectable and (2) with the present sensitivity of GC detectors, such sample volumes would by far exceed the total volumes of capillary columns. Thus sample preconcentration from a large-volume medium is a mandatory step prior to GC analysis. Uses of cold traps (116) and freezing the sample in an empty tube (117) or directly in a capillary column (10, 53) have been described. While the problems of water interference, solvent impurities, and sample changes have been observed with the first two approaches, on-column cryogenic concentration is technically difficult with large volumes of air samples.

At present, the best approach to concentrating trace organics from large volumes of air appears to be the use of a small sorbent column. Sufficient time to concentrate the compounds of interest is allowed by passing the dilute gas-phase sample through. The organics are subsequently recovered by either washing with a small amount of liquid or thermal desorption into a capillary column.

Choice of sorptive materials for the preparation of concentration precolumn could be critical. Either adsorption or partitioning effects can be utilized in such a concentration step. The underlying physicochemical phenomena were extensively investigated by Novak et al. (118–120). Based on the knowledge of solute retention behavior, absolute quantitative measurements are feasible for a limited number of components. However, the validity of these considerations for complex mixtures and capillary chromatography is very approximate (121) due to the complexity of multicomponent adsorption processes on these adsorbents (e.g., displacement phenomena and sorption kinetics). The possibilities and limitations of these sampling techniques have been discussed (121).

Classical adsorbents such as silica, alumina, or charcoal have somewhat limited utility as preconcentration media because of their excessive surface activity, irreversible adsorption of polar solutes, and the corresponding deactivation problems. In addition, the first two adsorbents exhibit a very strong affinity for water, which is a relatively major component of air samples. Whenever preconcentration and a following quantitative transfer of volatile organics into a gas chromatograph through thermal desorption are sought, with the least interference from atmospheric water, porous polymeric packings have proved to be materials of choice (112–124). Among them, 2,6-diphenyl-p-phenylene oxide (now widely known under the commercial name Tenax GC) has unusually high thermal stability and inertness toward most classes of organic compounds (121, 123, 124). Graphitized carbon black has also been used as a concentration medium in air pollution studies (125). Evaluation of different adsorbents was recently performed (126).

An early example of the effective use of preconcentration of atmospheric samples on a short porous polymer precolumn is shown in Figure 10. This figure compares the chromatographic profiles of trace organics concentrated in the atmosphere of room (A) before and after someone smoked a marijuana cigarette (B) and a tobacco cigarette (C), demonstrating direct concentration of cannabinoids and nicotine from air (124).

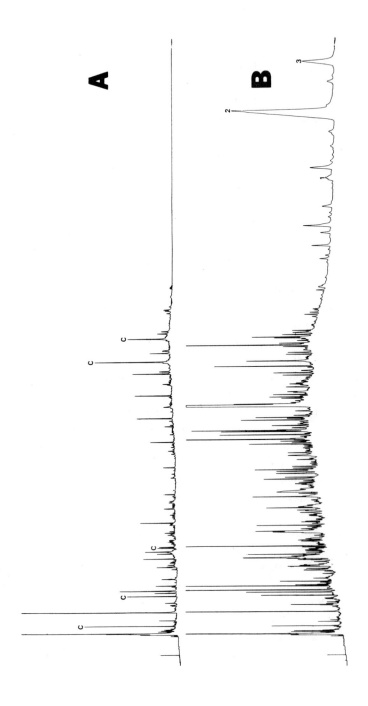

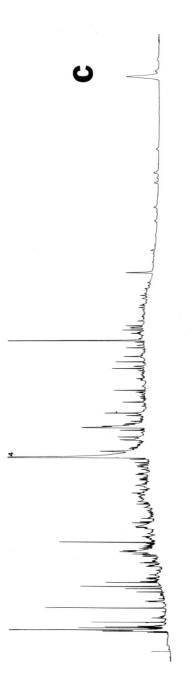

Figure 10. Chromatographic profiles obtained after a precolumn concentration and capillary GC of (A) background room atmosphere, (B) after a marijuana cigarette was smoked in the room, and (C) after a standard tobacco cigarette was smoked in the room. Conditions: 60 m × 0.4 mm i.d., glass capillary coated with SF-96 silicone fluid. Peaks: 1, cannabidiol; 2, Δ⁹-tetrahydrocannabinol; 3, cannabinol; 4, nicotine. Reproduced from Novotny and Lee (124) with permission of Birkhäuser Verlag, Basel.

97

Soon after the first applications of this sampling approach in early 1970s, it was clear that a coupling of this powerful concentration with highly efficient capillary columns would become popular in analyzing volatile complex mixtures. Numerous applications to air pollution studies have followed. Profiling hydrocarbon pollutants (127) and organic volatiles in Skylab 4 (128) and the detection of halogenated hydrocarbons (129) or a carcinogenic N-nitrosodimethylamine in ambient air at parts-per-trillion levels (130) may be quoted here as representative examples. Selectivity of the precolumn approach toward oxygenated species can be ascertained by the recently described nonvolatile coordination polymers of the lanthanides (131).

Another feasible approach to sample preconcentration consists of trapping the trace organics on a small amount of adsorbent with a relatively large surface area, followed by elution of the trapped material with a small volume of suitable solvent. For example, activated carbon was used for concentration of Zurich city air (132). After being rinsed with a small volume of CS_2, a sample aliquot was injected onto a glass capillary column that resolved well over 100 peaks. Mixture components were tentatively identified as aliphatic and aromatic hydrocarbons, chlorinated substances, and certain oxygen-containing compounds. As demonstrated earlier by Aue and Teli (133), certain siliceous materials with chemically bonded (nonextractable) stationary phases can also be effective in reaching similar goals.

Various organic pollutants may be encountered on the surface of airborne particulates. The particulate matter is collected by specialized devices on glass-fiber filters that are subsequently extracted with organic solvents. Although the analysis of concentrated extracts could aim at any of a variety of environmental pollutants, polycyclic aromatic hydrocarbons have been the most popular subject of analytical studies. Interest in this class of compounds is understandable because of their well-known carcinogenic properties. Recent contributions of capillary gas chromatography to this class of compounds are very significant. Many isomeric substances with profoundly different toxicological properties can now be resolved and quantitated. For example, polycyclic compounds with different positions of an alkyl group on the parent hydrocarbons can have different tumor-initiating activities: chrysene and 1-, 2-, 4-, and 6-methylchrysenes are potent carcinogens (103). Chromatographic resolution of these and similar isomers is now feasible (134).

A typical complexity of polycyclic aromatic mixtures extracted from airborne particulates is revealed in Figure 11, showing resolution of well over 100 components with a relatively short glass capillary column (73). It should be noted that relatively large molecules, such as coronene, are eluted from such columns at reasonably low temperatures. This is yet another distinct advantage of thin-film glass capillary columns. With further advances in surface deactivation and greater column stabilities (27, 31, 135), even larger molecules can now be eluted (136).

Although most studies have been directed toward "neutral" polycyclic aromatics at this date, the presence of various oxygenated aromatic species plus sulfur- and nitrogen-containing analogs to polycyclic aromatics in various environmental samples has been an important concern lately. Several such compounds are now suspected of having profound toxicological significance. Capillary GC and other analytical techniques used in the analysis of polycyclic aromatic compounds have recently been reviewed (134, 135).

A possible direct relationship between the content of polycyclic aromatic compounds in urban atmospheres and the incidence of lung cancer

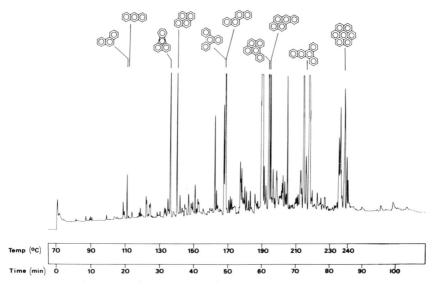

Figure 11. Capillary chromatogram of the total polynuclear aromatic hydrocarbon fraction of airborne particulates. Conditions: 11 m × 0.26 mm i.d., glass capillary column coated with SE-52 methylphenylsilicone stationary phase. Reprinted with permission from M. Lee et al., *Anal. Chem.* **48,** 1566 (1976). Copyright 1976 American Chemical Society.

(137) stimulates much interest in research toward rapid and reliable identification of different pollution sources and assessment of their health hazard. As shown by Bartle et al., (138) a simple procedure for particulate sample extraction, solvent partitioning, and capillary chromatography provides a fingerprinting technique capable of distinguishing samples from different urban areas. Further studies by Lee et al. (139) have extended this approach and demonstrated a typical formation of certain polycyclics due to different combustion conditions.

4.2. Water Pollution

Water contamination problems have received an increasing amount of attention. The current situation dictates that effectiveness of the previously used analytical methods for water quality evaluation be reevaluated. New standard methods need to be developed for an increasing amount of industrial chemicals. Our knowledge of the various mechanisms of different water treatment methods, the effects of the accumulation and biodegradation of certain industrial pollutants in waters with slower exchange rates (e.g., the Great Lakes on the North American Continent), the continued discharge of various pollutants, and repeated sanitary treatments of large rivers at different sites (e.g., the Mississippi River or the River Rhine in Europe) is frequently limited. In many such cases, a sample of polluted water may contain thousands of various chemicals present in different concentrations. Even though many such substances are hardly measurable by conventional chemist's tools, they may adversely effect the quality of drinking water or cause a serious health hazard to human beings and wildlife.

Although still in an early developing stage of application, capillary GC and its ancillary techniques may well be the solution to a problem of hitherto inadequate methodology in this area. To this date, capillary GC has been used on occasions in various applications to organic constituents in water, covering applications as diverse as fingerprinting oil spills (75), detecting pesticides in river water (15, 84), and identifying chlorinated substances in the city water supply of New Orleans (140) and in sewage (82). The series of papers by Grob (11–13) can now be considered the earliest systematic contribution to the analytical methodology of organic water pollutants using capillary columns.

It has now been shown in numerous studies that organics from various water sources can be effectively extracted and resolved with efficient capillary columns. Well over 100 compounds were identified (11–13) as hydrocarbons, chlorinated substances, terpenes, various industrial chemicals, and sulfur compounds (most likely breakdown products of bacterial processes). It has further been demonstrated that such organics can be identified at concentrations down to a few nanograms per liter, but the analyses are considered semiquantitative. High-resolution GC is indeed a very powerful method for comparative analyses of samples obtained from the same water source under different conditions, as evidenced by Figure 12 (12). Two competitive sampling procedures, gas-phase stripping with subsequent sample adsorption and a small-volume extraction, have been carefully compared (13), resulting in the conclusion that their relative merits for a given sample type must be considered individually.

Another approach to the analysis of volatile components of aqueous media is a combination of gas-phase stripping and concentration on the porous polymeric precolumns. As discussed in the previous section, organic porous polymers have low affinity for water, yet they trap effectively (but reversibly) a variety of organic compounds that can be thermally desorbed into a capillary column. Again, several applications in this area exist. Representative examples of this approach are the identification studies on river water (140, 141) and comparisons of volatile profiles of organics in drinking water subsequent to different treatments (142, 143).

Many aspects of capillary gas chromatography and related techniques are similar for both air and water pollution analysis. Even the substances of analytical interest may overlap (halocarbons, priority pollutants, polycyclic aromatic hydrocarbons, etc.). Consequently, there is little need to discuss these subjects widely here. However, it should be emphasized that the overall procedures for concentration of pollutants may differ substantially. For instance, more polar and less volatile pollutants may be recovered in water pollution studies using solvent extraction, adsorption on organic resins, ion-exchangers, and so on. Volatilization of such compounds can frequently be accomplished through the application of chemical derivatization techniques prior to capillary GC. Such techniques are employed widely in biochemical analysis, but less commonly in environmental studies. However, examples of the determination of phenols and acids in polluted water samples have been shown (144, 145). A wider utilization of derivatization techniques in the future may be anticipated.

It should be noted that while analyzing trace organics in water or air

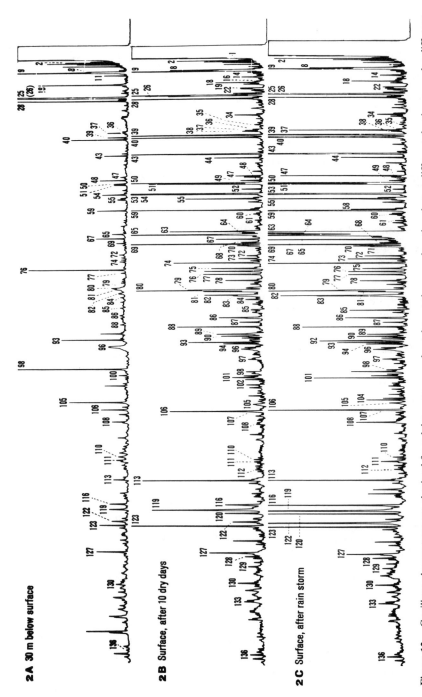

2A 30 m below surface

2B Surface, after 10 dry days

2C Surface, after rain storm

102

Figure 12. Capillary chromatograms obtained from lake water samples taken at the same location, but at different depths, or under different weather conditions. Chromatographic conditions: 120 m × 0.32 mm i.d., glass capillary column coated with a polypropylene glycol stationary phase, temperature-programmed from 25 to 175°C at 1.5°C/min. Reprinted from Grob and Grob (12) with permission of the Elsevier Publishing Company.

pollution studies, well-controlled and "clean" sampling techniques are extremely important. Contamination problems associated with solvent impurities and organic adsorbents must be minimized in order to take advantage of the "clean" capillary chromatographic systems that are currently available.

5.   RELATION OF CAPILLARY GC TO OTHER ANALYTICAL METHODS

No universal analytical method currently exists in the field of trace organic analysis. Thus no false expectations should be derived in our case. Capillary GC, like any other analytical technique, has its strengths and limitations. A proper placement of this technique in the overall analytical scheme and its combination with powerful related techniques can both enhance the strengths and reduce its limitations in environmental analytical chemistry. Thus we will now discuss briefly some complementary and competitive uses of other methods.

First of all, it is important to remember that GC covers sufficiently only substances that are relatively volatile. Unfortunately, many large and polar molecules have increasing technological and environmental significance. The methods for their determination have yet to be developed. With increasing molecular weight, the number of isomeric compounds to be resolved will increase proportionally. Although HPLC is fully capable of handling such large and polar molecules, significant improvements in column performance will be needed.

Some initial inquiries into the composition of nonvolatile organic mixtures are quite revealing. As shown in Figure 13, a high-molecular-weight fraction extracted from carbon black yields a chromatogram with very large polycyclic aromatic molecules (146), while using reversed-phased HPLC and fluorimetric detection. However, other mixtures of similar molecular weight range show very severe resolution problems. It should be remembered that capillary GC yields typical efficiencies between 10^5 and 10^6 theoretical plates, while 10^4 plates are seldom exceeded in the art of conventional HPLC.

Recent attempts to increase column efficiencies in HPLC through the development of microcolumn technology and miniaturized systems (147–149) have met with promising results. Various aspects of this new direc-

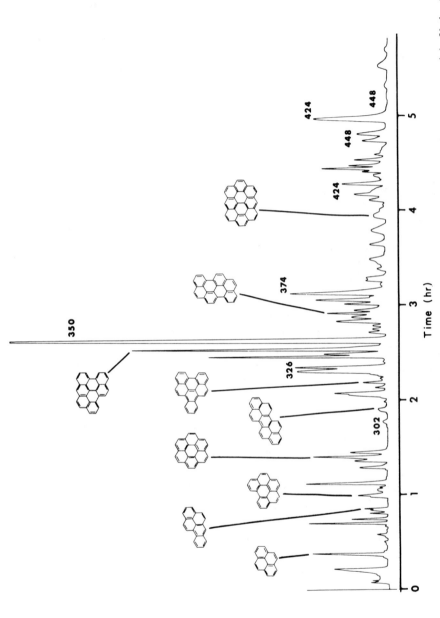

Figure 13. HPLC analysis of high-molecular-weight fraction extracted from carbon black. Conditions: 25 cm × 4.6 mm i.d., Vydac C$_{18}$ reversed-phase column. Gradient elution: 50:50 water–acetonitrile for 15 min, then to 100% acetonitrile for 70 min, to 100% ethyl acetate for 130 min, and to 100% methylene chloride during the last 60 min. Reprinted with permission from P. A. Peaden et al., (146) *Anal. Chem.* **52,** 2268 (1980) Copyright 1980 American Chemical Society.

tion, including detection techniques, have recently been reviewed (150). Capillary supercritical fluid chromatography (151, 152) and capillary zone electrophoresis (153) are two additional techniques that are likely to expand high-resolution capabilities into the area of nonvolatile mixtures. However, as these separation capabilities fully materialize, new ancillary techniques will increasingly be needed to meet structural elucidation tasks. While expanding into the area of larger molecules, the above-mentioned techniques will be truly complementary to the current utilization of capillary GC in environmental analytical chemistry.

The current HPLC techniques can also be of significant help in simplifying the mixture complexity prior to capillary GC. There are several reasons of this approach. Many environmental samples are so complex that even the best capillary columns frequently fail to resolve all mixture components. Sample fractionation by partition schemes merely provides a rough separation into compound classes. In contrast, HPLC serves as an effective fractionation approach. Figure 14 demonstrates an example of this: The basic (aza-arene) fraction of cannabis smoke was fractionated by HPLC with an aminosilane column (as discussed earlier), while the selected "cuts" were concentrated and injected into an efficient glass capillary column, resulting in complex chromatograms and little overlap (154, 155). Somewhat similar fractionations were also reported for polycyclic aromatic hydrocarbon mixtures from airborne particulates (73) and phenylurea herbicides (156).

Current HPLC techniques can be competitive to capillary GC whenever sample volatility is marginal, mixture complexity is not excessive, or when a finite number of components must be selectively detected (e.g., with fluorimetric or electrochemical detectors). Analysis of polycyclic aromatic compounds (134) is a case in point where both analytical techniques, depending on a particular problem or a personal preference for one method over the other, continue providing satisfactory results.

Any analytical technique is being continuously challenged by newer developments, and so is capillary GC. Once the chemical composition of complex mixtures becomes understood through the use of multicomponent analytical methods (such as capillary GC or HPLC) and important environmental criteria are assigned to a selected number of compounds, future efforts should be directed toward methodological simplification. However, this will not diminish the value of chromatographic methods for screening purposes. These will be continuously needed due to the

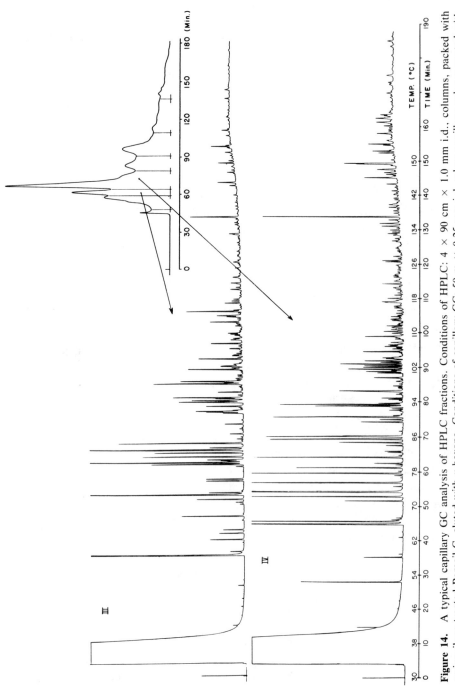

Figure 14. A typical capillary GC analysis of HPLC fractions. Conditions of HPLC: 4×90 cm $\times$ 1.0 mm i.d., columns, packed with aminosilane-treated Porasil C., eluted with *n*-hexane. Conditions of capillary GC: 50 m $\times$ 0.25 mm i.d., glass capillary column coated with UCON 50-HB-2000. Reprinted with permission from F. Merli et al., *Anal. Chem.* **53**, 1929 (1981). Copyright 1981 American Chemical Society.

106

occurrence of new chemicals in the environment, process modifications, and technological innovations that may create pollution.

Development of new spectroscopic techniques will undoubtedly influence high-resolution chromatographic methods in future. Interestingly, with respect to a recent emphasis on MS/MS techniques (157, 158), some view these developments as an effective replacement for GC/MS, while others argue that chromatography and MS/MS will actually benefit each other.

REFERENCES

1. M. J. E. Golay, U. S. Patent 2,920,478 (1960).
2. I. Halasz and C. Horvath, *Nature* **197,** 71 (1963).
3. I. Halasz and C. Horvath, *Anal. Chem.* **35,** 499 (1963).
4. D. H. Desty and A. Goldup, in *Gas Chromatography*, R. P. W. Scott, Ed., Butterworths, London, 1960, p. 162.
5. D. H. Desty, *Adv. Chromatogr.* **1,** 199 (1965).
6. K. Grob, *Helv. Chim. Acta* **48,** 1362 (1965).
7. K. Grob, *Helv. Chim. Acta* **51,** 718 (1968).
8. M. Novotny and K. Tesarik, *Chromatographia* **1,** 332 (1968).
9. K. Tesarik and M. Novotny, in *Gas-Chromatographie,* H. G. Struppe, Ed., Akademie-Verlag, Berlin, 1968, p. 575.
10. K. D. Bartle, L. Bergstedt, M. Novotny, and G. Widmark, *J. Chromatogr.* **45,** 256 (1969).
11. K. Grob, *J. Chromatogr.* **84,** 255 (1973).
12. K. Grob and G. Grob, *J. Chromatogr.* **90,** 303 (1974).
13. K. Grob, K. Grob, Jr., and G. Grob, *J. Chromatogr.* **106,** 299 (1975).
14. R. Dandeneau and E. H. Zerenner, *J. High Resoln. Chromatogr.* **2,** 351 (1979).
15. D. H. Desty, J. N. Haresnape, and B. H. F. Whyman, *Anal. Chem.* **32,** 302 (1960).
16. F. Farré-Rius, J. Henniker, and G. Guiochon, *Nature* **196,** 63 (1962).
17. A. Liberti, in *Gas Chromatography*, A. B. Littlewood, Ed., Elsevier, Amsterdam, 1967, p. 95.
18. G. Alexander and G. A. F. M. Rutten, *J. Chromatogr.* **99,** 81 (1974).
19. J. D. Schieke, N. R. Comins, and V. Pretorius, *J. Chromatogr.* **112,** 97 (1975).
20. F. I. Onuska, M. E. Comba, T. Bistricki, and R. J. Wilkinson, *J. Chromatogr.* **142,** 117 (1977).

21. K. Grob and G. Grob, *J. Chromatogr.* **125**, 471 (1976).
22. K. D. Bartle and M. Novotny, *J. Chromatogr.* **94**, 35 (1974).
23. M. Novotny and K. Grohmann, *J. Chromatogr.* **84**, 167 (1973).
24. A. M. Filbert and M. L. Hair, *J. Gas Chromatogr.* **6**, 218 (1968).
25. J. J. Franken and M. M. F. Trijbels, *J. Chromatogr.* **91**, 425 (1974).
26. K. Grob and G. Grob, *Chromatographia* **4**, 422 (1971).
27: M. L. Lee, D. L. Vassilaros, L. V. Phillips, D. M. Hercules, H. Azumaya, J. W. Jorgenson, M. P. Maskarinec, and M. Novotny, *Anal. Lett.* **12**, 191 (1979).
28. B. W. Wright, M. L. Lee, S. W. Graham, L. V. Phillips, and D. M. Hercules, *J. Chromatogr.* **199**, 355 (1980).
29. W. A. Aue, C. R. Hastings, and S. Kapila, *J. Chromatogr.* **77**, 299 (1973).
30. D. A. Cronin, *J. Chromatogr.* **97**, 263 (1974).
31. L. Blomberg and T. Wännman, *J. Chromatogr.* **148**, 379 (1978).
32. L. D. Metcalfe and R. J. Martin, *Anal. Chem.* **39**, 1204 (1967).
33. G. A. F. M. Rutten and J. A. Luyten, *J. Chromatogr.* **74**, 177 (1972).
34. T. Welsch, W. Engewald, and C. Klaucke, *Chromatographia* **10**, 22 (1977).
35. K. Grob, G. Grob, and K. Grob, Jr., *J. High Resoln. Chromatogr.* **2**, 31 (1979).
36. L. Blomberg, K. Markides, and T. Wännman, in *Capillary Chromatography*, R. E. Kaiser, Ed., Hüthig, Heidelberg, 1981.
37. C. Madani, E. M. Chambaz, M. Rigaud, J. Durand, and P. Chebroux, *J. Chromatogr.* **126**, 161 (1976).
38. L. Blomberg and T. Wännman, *J. Chromatogr.* **186**, 159 (1979).
39. K. Grob, G. Grob, and K. Grob, Jr., *J. Chromatogr.* **211**, 243 (1981).
40. M. Novotny, L. Blomberg, and K. D. Bartle, *J. Chromatogr. Sci.* **8**, 390 (1970).
41. K. D. Bartle, *Anal. Chem.* **45**, 1831 (1973).
42. G. Schomburg, H. Husmann, and F. Weeke, *J. Chromatogr.* **99**, 63 (1974).
43. J. Bouche and M. Verzele, *J. Gas Chromatogr.* **6**, 501 (1968).
44. E. L. Ilkova and E. A. Mistryukov. *J. Chromatogr. Sci.* **9**, 569 (1971).
45. M. L. Lee and B. W. Wright, *J. Chromatogr.* **184**, 235 (1980).
46. W. Jennings, *Gas Chromatographay with Glass Capillary Columns*, 2nd ed., Academic, New York, 1980
47. J. C. Giddings, in *Gas Chromatography*, A. Goldup, Ed., The Institute of Petroleum, London, 1965, p. 3.
48. G. Guiochon, *Anal. Chem.* **50**, 1812 (1978).
49. C. P. M. Schutjes, E. A. Vermeer, J. A. Rijks, and C. A. Cramers, in *Capillary Chromatography*, R. E. Kaiser, Ed., Hüthig, Heidelberg, 1981, p. 687.

50. K. Grob, Jr., G. Grob, and K. Grob, *J. Chromatogr.* **156,** 1 (1978).

51. J. J. Franken and G. A. F. M. Rutten, in *Gas Chromatography*, S. G. Perry, Ed., Applied Science Publishers, Barking, U.K., 1973, p. 75.

52. K. Grob and G. Grob, *J. Chromatogr. Sci.* **7,** 584 (1969).

53. D. R. Rushneck, *J. Gas Chromatogr.* **3,** 319 (1965).

54. K. Grob and K. Grob, Jr., *J. Chromatogr.* **94,** 53 (1974).

55. F. J. Yang, A. C. Brown III, and S. P. Cram, *J. Chromatogr.* **158,** 91 (1978).

56. K. Grob and K. Grob, Jr., *J. Chromatogr.* **151,** 311 (1978).

57. M. Galli, S. Trestianu, and K. Grob, Jr., *J. High Resoln. Chromatogr.* **2,** 366 (1979).

58. M. Galli and S. Trestianu, *J. Chromatogr.* **203,** 193 (1981).

59. P. M. J. van den Berg and T. P. H. Cox, *Chromatographia* **5,** 301 (1972).

60. M. Novotny and R. Farlow, *J. Chromatogr.* **103,** 1 (1975).

61. G. Schomburg, H. Husmann, and F. Weeke, *J. Chromatogr.* **112,** 205 (1975).

62. M. L. Lee, K. D. Bartle, and M. Novotny, *Anal. Chem.* **47,** 540 (1975).

63. A. Karmen and L. Giuffrida, *Nature* **201,** 1204 (1964).

64. V. V. Brazhnikov, M. V. Gur'ev, and K. I. Sakodynsky, *Chromatogr. Rev.* **12,** 1 (1970).

65. M. Krejci and M. Dressler, *Chromatogr. Rev.* **13,** 1 (1970).

66. B. Kolb and J. Bischoff, *J. Chromatogr. Sci.* **12,** 625 (1974).

67. M. J. Hartigan, J. E. Purcell, M. Novotny, M. L. McConnell, and M. L. Lee, *J. Chromatogr.* **99,** 339 (1974).

68. S. G. Wakeham, *Environ. Sci. Technol.* **13,** 1119 (1979).

69. N. P. Buu-Hoi, F. Zajdela, O. Roussel, and L. Petit, *Bull. Cancer* **52,** 49 (1965).

70. F. M. Zado and R. S. Juvet, *Anal. Chem.* **38,** 569 (1966).

71. S. S. Brody and J. E. Chaney, *J. Gas Chromatogr.* **4,** 42 (1966).

72. M. L. Lee, Ph.D. thesis, Indiana University, 1975.

73. M. L. Lee, M. Novotny, and K. D. Bartle, *Anal. Chem.* **48,** 1566 (1976).

74. M. L. Lee and R. A. Hites, *Anal. Chem.* **48,** 1890 (1976).

75. E. R. Adlard, L. F. Creasar, and P. H. D. Mathews, *Anal. Chem.* **44,** 64 (1972).

76. W. E. Rupprecht and T. R. Phillips, *Anal. Chim. Acta* **47,** 439 (1969).

77. S. G. Perry and F. W. G. Carter, in *Gas Chromatography*, N. Stock and S. G. Perry, Eds., The Institute of Petroleum, London, 1971, p. 381.

78. J. E. Lovelock, *J. Chromatogr.* **99,** 3 (1974).

79. J. E. Lovelock and A. J. Watson, *J. Chromatogr.* **158,** 123 (1978).

80. P. Deveaux and G. Guiochon, *Chromatographia* **2,** 151 (1969).

81. D. C. Fenimore, P. R. Loy, and A. Zlatkis, *Anal. Chem.* **43**, 1972 (1971).

82. K. Grob, *Chromatographia* **8**, 423 (1975).

83. K. Grob and G. Grob, *J. Chromatogr. Sci.* **8**, 635 (1970).

84. J. J. Franken and H. L. Vader, *Chromatographia* **6**, 22 (1973).

85. M. Suzuki, Y. Yamato, and T. Watanabe, *Environ. Sci. Toxicol.* **11**, 1109 (1977).

86. A. Björseth and G. Eklund, *J. High Resoln. Chromatogr.* **2**, 22 (1979).

87. E. P. Grimsrud and D. A. Miller, *Anal. Chem.* **50**, 1141 (1978).

88. R. E. Sievers, M. P. Phillips, R. M. Barkley, M. A. Wizner, M. J. Bollinger, R. S. Hutte, and F. C. Fehsenfeld, *J. Chromatogr.* **186**, 3 (1979).

89. J. E. Lovelock, *Anal. Chem.* **33**, 163 (1961).

90. J. N. Driscoll, J. Ford, J. F. Jaramillo, and E. T. Gruber, *J. Chromatogr.* **158**, 171 (1978).

91. J. N. Driscoll, J. Ford, L. F. Jaramillo, J. H. Becker, G. Hewitt, J. K. Marshall, and F. Onishuk, *Am. Lab.* **10**, 137 (1978).

92. J. F. Jaramillo and J. N. Driscoll, *J. High Resoln. Chromatogr.* **2**, 536 (1979).

93. G. W. Price, D. C. Fenimore, P. G. Simmonds, and A. Zlatkis, *Anal. Chem.* **40**, 541 (1968).

94. W. Kaye, *Anal. Chem.* **34**, 287 (1962).

95. H. P. Burchfield, R. J. Wheeler, and J. B. Bernos, *Anal. Chem.* **43**, 1972 (1971).

96. P. J. Freed and L. R. Faulkner, *Anal. Chem.* **44**, 1194 (1972).

97. R. P. Cooney and J. D. Winefordner, *Anal. Chem.* **49**, 1057 (1977).

98. H. H. Hausdorff, *J. Chromatogr.* **134**, 131 (1977).

99. M. Novotny, F. J. Schwende, M. J. Hartigan, and J. E. Purcell, *Anal. Chem.* **52**, 736 (1980).

100. D. Kuehl and P. R. Griffiths, *Anal. Chem.* **52**, 1394 (1980).

101. D. M. Hembree, A. A. Garrison, R. A. Crocombe, R. A. Yokley, E. L. Wehry, and G. Mamantov, *Anal. Chem.* **53**, 1783 (1981).

102. R. P. Cooney, T. Vo-Dinh, and J. D. Winefordner, *Anal. Chim. Acta* **89**, 9 (1977).

103. D. Hoffmann, W. E. Bondinell, and E. L. Wynder, *Science* **183**, 215 (1974).

104. M. L. Lee, D. L. Vassilaros, C. M. White, and M. Novotny, *Anal. Chem.* **51**, 768 (1979).

105. M. Novotny, R. Kump, F. Merli, and L. J. Todd, *Anal. Chem.* **52**, 401 (1980).

106. M. L. Lee, D. L. Vassilaros, W. S. Pipkin, and W. L. Sorenson, in "Trace Organic Analysis: A New Frontier in Analytical Chemistry," National Bureau of Standards Special Publication 519, H. S. Hertz and S. N. Chesler, Eds., U. S. Government Printing Office, Washington, D.C., 1979, p. 731.

107. E. C. Horning and M. G. Horning, *J. Chromatogr. Sci.* **9**, 129 (1971).
108. M. Novotny, M. L. McConnell, M. L. Lee, and R. Farlow, *Clin. Chem.* **20**, 1105 (1974).
109. M. L. McConnell, G. Rhodes, U. Watson, and M. Novotny, *J. Chromatogr.* **162**, 495 (1979).
110. G. Rhodes, M. Miller, M. L. McConnell, and M. Novotny, *Clin. Chem.* **27**, 580 (1981).
111. A. Zlatkis, C. F. Poole, R. Brazell, K. Y. Lee, F. F. Hsu, and S. Singhawangcha, *Analyst (London)* **106**, 352 (1981).
112. E. Jellum, I. Björnson, R. Nesbakken, E. Johansson, and S. Wold, *J. Chromatogr.* **217**, 2313 (1981).
113. M. L. McConnell and M. Novotny, *J. Chromatogr.* **112**, 559 (1975).
114. H. Hrivnac, W. Frischknecht, and M. Cechova, *Anal. Chem.* **48**, 937 (1976).
115. H. A. Clark and P. C. Jurs, *Anal. Chem.* **47**, 374 (1975).
116. H. M. Liebich, W. A. Koening, and E. Bayer, *J. Chromatogr. Sci.* **8**, 527 (1970).
117. R. A. Flath, R. R. Forrey, and R. Teranishi, *J. Food Sci.* **34**, 382 (1979).
118. J. Novak, V. Vasak, and J. Janak, *Anal. Chem.* **37**, 660 (1965).
119. J. Gelbicova-Ruzickova, J. Novak, and J. Janak, *J. Chromatogr.* **64**, 15 (1972).
120. J. Novak, J. Janak, and J. Golias, in "Trace Organic Analysis: A New Frontier in Analytical Chemistry," National Bureau of Standards Special Publication 519, H. S. Hertz and S. N. Chesler, Eds., U. S. Government Printing Office, Washington, D.C., p. 739.
121. M. Novotny, M. L. Lee, and K. D. Bartle, *Chromatographia* **7**, 333 (1974).
122. A. Dravnieks and A. O'Donnell, *J. Agric. Food Chem.* **19**, 1049 (1971).
123. A. Zlatkis, H. A. Lichtenstein, and A. Tishbee, *Chromatographia* **6**, 67 (1973).
124. M. Novotny and M. L. Lee, *Experientia* **29**, 1038 (1973).
125. P. Ciccioli, G. Bertoni, E. Brancaleoni, R. Fratarcangeli, and F. Bruner, *J. Chromatogr.* **126**, 757 (1976).
126. J. E. Bunch and E. D. Pellizzari, *J. Chromatogr.* **186**, 811 (1979).
127. W. Bertsch, R. C. Chang, and A. Zlatkis, *J. Chromatogr. Sci.* **12**, 175 (1974).
128. W. Bertsch, A. Zlatkis, H. M. Liebich, and H. J. Schneider, *J. Chromatogr.* **99**, 673 (1974).
129. E. D. Pellizzari, J. E. Bunch, R. E. Berkley, and J. McRae, *Anal. Chem.* **48**, 803 (1976).
130. E. D. Pellizzari, J. E. Bunch, J. T. Bursey, R. E. Berkley, E. Sawicki, and K. Krost, *Anal. Lett.* **9**, 579 (1976).
131. J. E. Picker and R. E. Sievers, *J. Chromatogr.* **217**, 275 (1981).

132. K. Grob and G. Grob, *J. Chromatogr.* **62**, 1 (1971).

133. W. A. Aue and P. M. Teli, *J. Chromatogr.* **62**, 15 (1971).

134. M. L. Lee, M. Novotny, and K. D. Bartle, *Analytical Chemistry of Polycyclic Aromatic Compounds*, Academic, New York, 1981.

135. M. L. Lee and B. W. Wright, *J. Chromatogr. Sci.* **18**, 345 (1980).

136. Y. Hirata, M. Novotny, P. A. Peaden, and M. L. Lee, *Anal. Chim. Acta* **127**, 55 (1981).

137. L. B. Lave and E. P. Seskin, *Science* **169**, 723 (1970).

138. K. D. Bartle, M. L. Lee, and M. Novotny, *Int. J. Environ. Anal. Chem.* **3**, 349 (1974).

139. M. L. Lee, G. P. Prado, J. B. Howard, and R. A. Hites, *Biomed. Mass Spectrom.* **4**, 182 (1977).

140. B. Dowty, D. Carlisle, J. L. Laseter, and J. Storer, *Science* **187**, 75 (1975).

141. W. Bertsch, E. Anderson, and G. Holzer, *J. Chromatogr.* **112**, 701 (1975).

142. M. G. Black, W. R. Rehg, R. E. Sievers, and J. J. Brooks, *J. Chromatogr.* **142**, 809 (1977).

143. M. Thomason, M. Shoults, W. Bertsch, and G. Holzer, *J. Chromatogr.* **158**, 437 (1978).

144. R. Gloor and H. Leidner, *Chromatographia* **9**, 618 (1976).

145. J. J. Richard, C. D. Chriswell, and J. S. Fritz, *J. Chromatogr.* **199**, 143 (1980).

146. P. A. Peaden, M. L. Lee, Y. Hirata, and M. Novotny, *Anal. Chem.* **52**, 2268 (1980).

147. R. P. W. Scott, *Analyst (London)* **103**, 37 (1978).

148. T. Tsuda and M. Novotny, *Anal. Chem.* **50**, 271 (1978).

149. K. Hibi, T. Tsuda, T. Takeuchi, T. Nakanishi, and D. Ishii, *J. Chromatogr.* **175**, 105 (1979).

150. M. Novotny, *Anal. Chem.* **53**, 1294A (1981).

151. M. Novotny, S. R. Springston, P. A. Peaden, J. C. Fjeldstedt, and M. L. Lee, *Anal. Chem.* **53**, 407A (1981).

152. S. R. Springston and M. Novotny, *Chromatographia* **14**, 679 (1981).

153. J. W. Jorgenson and K. D. Lukacs, *Anal. Chem.* **53**, 1298 (1981).

154. F. Merli, M. Novotny, and M. L. Lee, *J. Chromatogr.* **199**, 371 (1980).

155. F. Merli, D. Wiesler, M. P. Maskarinec, M. Novotny, D. L. Vassilaros, and M. L. Lee, *Anal. Chem.* **53**, 1929 (1981).

156. A. de Kok, A. A. van der Kooij, M. Linnekamp, Y. J. Vos, R. W. Frei, and U. A. T. Brinkman, in *Capillary Chromatography*, R. E. Kaiser, Ed., Hüthig, Heidelberg, 1981, p. 53.

157. R. W. Kondrat and R. G. Cooks, *Anal. Chem.* **50**, 81A (1978).

158. R. A. Yost and C. G. Enke, *Anal. Chem.* **51**, 1251A (1979).

RECENT ADVANCES IN
GAS CHROMATOGRAPHY/MASS SPECTROMETRY
IN ENVIRONMENTAL ANALYSIS

PHILIP W. RYAN

Systems, Science and Software
P.O. Box 1620
La Jolla, California 92038

1. INTRODUCTION

Since its inception in 1957, the coupled gas chromatograph/mass spectrometer instrument (GC/MS) has become one of the indispensable tools of analytical chemistry, particularly of organic environmental chemistry. The reasons are clear to anyone who has ever contemplated studying the chemistry of a typical environmental sample. Complexity and diversity are characteristic of the samples collected by an environmental chemist, and the analytical task is frequently to identify and measure one or more chemical constituents among the hundreds, thousands, or more present. The chemist may be attempting to determine whether some particular compound is present at parts-per-billion concentrations, or he or she may be simply out to discover what is present at the parts-per-billion level. To attack such a problem, a highly specific detection scheme is required, one which provides such distinctive information that the detected compound's identity is established beyond reasonable doubt. Furthermore,

113

such information must be obtained for each substance present and with sufficient sensitivity that even trace components can be observed.

Although no single analytical technique can ever live up to such a requirement, mass spectrometry comes closer than most. The information contained in a mass spectrum is often sufficient to unambiguously identify a compound; it is always sufficient to severely limit the possibilities and to reveal features of molecular structure. Any compound with a vapor pressure greater than 10^{-7} mm Hg at 250°C can be observed, and typically no more than a few nanograms of material are required. In special cases, subpicogram amounts can be detected, giving mass spectrometry a sensitivity as good as any other means of detecting organic molecules.

If a mixture of compounds is introduced into a mass spectrometer, the resulting mixture spectrum still contains the full set of structural information for each component, but only in the simplest cases is it feasible to sort that information into sets associated with discrete components. Seldom does any sample of environmental interest constitute one of the simple cases. With the addition of a gas chromatograph, the mass spectral information is spread out along a time axis and, given sufficient time (chromatographic) resolution, the data may be rendered interpretable. It is from this combination of a detector that provides specific information about a pure compound with a means of separating complex samples so that pure components can be observed that the analytical power of GC/MS is derived. A fortunate compatibility between separator and detector is that both operate on samples with similar physical characteristics. Any substance that can be eluted from a GC column possesses a volatility sufficient for mass spectrometer analysis. Furthermore, both techniques are set up to accept approximately the same sample amounts, nanograms to micrograms, so that overloading and detection capability limitations are roughly the same. One fortunate result is that the GC acts as a filter that minimizes the most common sources of MS deterioration: too much sample in the ion source and contamination by substances too involatile to be pumped away quickly. This filtering action of the GC is almost as important as its separation powers in making the analytical capabilities of MS available to environmental chemistry.

For GC/MS, a data system is far more than a mere convenience: It is an essential part that greatly amplifies the analytical power of the instrument. This is a consequence of the overwhelming amount of information that can be generated during a GC/MS run. Quite apart from its sophis-

ticated applications in library searching, structure elucidation, data enhancement, and so on, a data system is indispensable simply to record,
store, and recall that information.

2. INSTRUMENTATION

All four essential components of a modern GC/MS system—gas chromatograph, interface, mass spectrometer, and data system—have experienced significant development in the past few years that has vastly increased the usefulness of this instrument in environmental analysis.

With the exception of detector development, essentially any progress
in gas chromatography is progress in GC/MS. One advance that is perhaps
more important to GC/MS than to GC in general is the recent improvement
in several of the widely used liquid phases to minimize column bleed.
Bleed is the release of degradation products or impurities, such as lower
polymers, from the bulk stationary phase of the GC column. When these
substances enter the spectrometer ion source, they not only contribute
an immediate mass spectral background (which varies during the run) but
also condense on critical surfaces and degrade the spectrometer's performance. Contamination due to bleed is often the most important cause
of dirty ion source with its resultant down time or poor quality spectra.
One example of bleed reduction is in the recent introduction of high-
molecular-weight dimethyl silicone liquids such as OV-101 or SP-2100 to
replace such bleeders as DC-200 (1). A second example is found in the
development of liquid crystal phases used to separate geometric isomers
of polynuclear aromatic hydrocarbons. Severe bleed problems made the
earlier liquid crystals unsuitable for GC/MS, but more recently developed
phases have minimized the problem (2). This is a particularly important
case for GC/MS because such geometric isomers give identical EI and
conventional chemical ionization (CI) spectra, so that chromatographic
separation is essential to identifying them.

The most dramatic recent advance in environmental GC/MS, as in
environmental GC, is the development of high-resolution capillary columns. This technique exploded into prominence quite apart from any
association with mass spectrometry with its phenomenally successful application around 1970 to complex environmental samples. Just after capillary columns had become an accepted GC/MS accessory, another tech-

nological breakthrough made fused silica capillary columns available. The superior performance of fused silica columns combined with the simplicity of using them has given every GC/MS easy access to the power of high-resolution chromatography. By revealing ten components where before there had been one, however, high-resolution chromatography has added at least as much confusion as clarification to many issues. The use of capillary columns is reviewed in detail by Novotny (3). The benefits of simplifying mixture spectra are obvious, but no less significant is the complementary power of a mass spectrometer in making sense out of the thicket of peaks into which a capillary column can transform a sample.

Capillary chromatography and GC/MS are techniques whose combination ought to yield benefits beyond the sum of the individual parts. The realization of those benefits has demanded mass spectrometer performance features that had not been previously considered important in instrument design.

The interface between GC and MS is needed to transfer sample from the high-pressure operating region of the GC ($\sim$700 Torr of carrier gas) to the high vacuum condition of the spectrometer ($\sim 10^{-5}$ Torr). This transfer must occur without losing too much sample and while preserving the spatial separation of eluted compounds. A variety of molecular separators is available for this purpose. In contrast to helium, many organic molecules are readily soluble in certain silicones that can be used as membranes to selectively pass soluble molecules from the GC effluent stream to the spectrometer. Helium atoms are able to diffuse relatively rapidly through a porous material, and can therefore be selectively removed from the gas stream between the GC and MS. Today, the most commonly used device is the jet separator, where carrier and entrained sample gas expand through a nozzle into a lower pressure chamber. More massive sample molecules tend to travel along the jet axis and enter a capillary passage leading to the mass spectrometer ion source, while lighter helium is likely to possess so much radial velocity that most of it misses the second orifice and is pumped away. These devices can be better than 50% efficient in passing sample, while eliminating 90% of the helium. In practice, separation efficiency is strongly dependent on the nature (polarity or mass) of the sample and on such operating conditions as temperature and carrier flow rate. Dependent on the same parameters is some loss in chromatographic resolution. Careful operation, however, can minimize this loss, and jet separators have been successfully used even with wall-coated open tubular (WCOT) columns (3).

It is becoming increasingly common to dispense with molecular separators entirely and run the GC effluent directly into the MS ion source. When using WCOT columns, carrier flow is usually less than 1 mL/min, and most mass spectrometer pumping systems can maintain sufficient vacuum under these conditions. Support-coated open tubular (SCOT) columns (3) typically operate with a carrier flow of 2–5 mL/min. This presents no problem for most modern instruments, which are designed with high capacity differential pumping arrangements in order to cope with the large gas flows associated with chemical ionization mass spectrometry. However, many older instruments cannot tolerate flows above 1 mL/min. With chemical ionization sources, there is no need for any separator and combined carrier/reagent gas flows of 20 mL/min are the rule.

With direct GC/MS coupling, all parts of the transfer lines feel the spectrometer pumping system, and GC effluent immediately enters a low-pressure region on leaving the column. Therefore, it travels from column to ion source with very great linear velocity, and postcolumn dead volume, often a detriment to resolution in capillary chromatography, is essentially eliminated. Unless some sort of flow restriction is added, there is also a partial vacuum created in the last part of the column. This can lead to some loss of chromatographic resolution, and it distorts the chromatogram so that absolute retention times are not the same as measured with other detectors. Retention indices and relative retention times, however, retain their usefulness (4). Figure 1 compares WCOT chromatograms obtained from the same sample with conventional FID detection and with a directly coupled GC/MS instrument. In spite of the capillary restrictor used in this case, minor changes in retention time of early peaks are noticeable, but there is no difficulty in correlating the two sets of peaks. Fused silica capillary columns have essentially put an end to the historical difficulties of GC/MS interfacing. These columns can be run right through the GC oven wall, through any convenient transfer oven, and into the MS. Vacuum seals can be made directly at the column with conventional graphite ferrules, and column flexibility means that the fragility and the critical alignment difficulties that have plagued coupling schemes with other kinds of columns are no longer serious. With the column exit positioned within millimeters of the electron beam, the direct

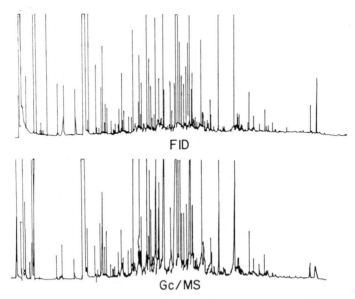

Figure 1. Comparison of WCOT chromatograms recorded with conventional flame ionization detector and with directly coupled GC/MS. Reprinted with permission from F. A. Thome and G. W. Young, *Anal. Chem.* **48**, 1423 (1976). Copyright 1976 American Chemical Society.

coupled fused silica column also provides maximum transfer efficiency with minimum dead volume.

Development of mass spectrometers for use in GC/MS has been strongly influenced by two trends of importance to the environmental chemist. These are the demands for capillary GC/MS instruments and the increasing popularity of chemical ionization mass spectrometry.

For a number of reasons, including the desirability of signal averaging, ease of peak detection and recording, and magnet hysteresis problems, mass spectrometers have traditionally been scanned rather slowly. In packed column GC/MS with repetitive scanning, it is usually sufficient to scan once every 5 or 10 sec in order to obtain several spectra during the elution of a GC peak. From a capillary column, however, peak widths of a few seconds are common. If the scan time is not several times smaller than the peak width, ion currents due to sample will change during the scan and spectra will be seriously distorted replicas of the true sample spectrum.

Instruments using quadrupole mass analyzers had the easier time adapting to capillary scan speed requirements because only electric fields had to be changed during a scan and reset times were negligible. Even quadrupole instruments designed for packed column GC/MS were capable of scanning at 300 amu/sec, which is fast enough for most WCOT chromatography and not far below the 1000 amu/sec upper limit of newer instruments designed with capillary columns in mind.

Magnetic sector mass analyzers must cope with hysteresis, the sluggish, nonlinear response of magnetic field to changing the potential drop across the magnet coils. Hysteresis limits the rate at which magnetic field can be changed reproducibly to focus different masses and also determines the time for resetting magnetic field to begin a new scan, the "flyback" time. Design advances in mass spectrometer magnets have recently enabled these instruments to perform as well as the quadrupoles in terms of scan speed. Scan rates faster than 0.5 sec/decade with flyback times of less than 0.5 sec are now commercially available.

These scan speeds approach the useful limit imposed by statistical considerations. The ultimate ability of a mass spectrometer to detect a mass peak is set by the signal-to-noise ratio at the detector. Noise levels depend on many factors ranging from electron multiplier shot noise to instrument contamination. Signal levels depend primarily on the ability of the instrument to convert molecules in the source region into ions at the detector. This figure is usually on the order of 10^{-9} C/μg of sample or about 10^7 ions/ng. These ions, of course, are distributed over the mass spectrum and also over the elution time of the GC peak so that the ion current at the maximum of a 5-sec GC peak containing 1.0 ng at a mass fragment which is 1% of total ionization is about 2.3×10^4 ions/sec. If 10 ions is the minimum average number to give a statistically recognizable signal for a typical noise level, then the sampling interval of the data system must be 4.3×10^{-4} sec. If 10 points are used to define a mass peak, the scan speed can not be more than 230 amu/sec, and even at this rate fragments less than 1% of total ionization will probably not be detected.

The use of capillary columns does help in two ways to increase GC/MS sensitivity over packed column use. By concentrating the nanogram of material in a 2-sec peak rather than the 10 or 20 for packed columns, the peak sample ion current is increased. At the same time, noise levels can be less because there are fewer unresolved potentially interfering compounds to contribute background ion current.

An interesting corollary of this signal-to-noise sensitivity limitation is the observation that quadrupole instruments often have a relatively high background current due to stray ions and energetic neutrals. As a result, the higher resolution, more background-free magnetic instruments can exhibit a superior practical sensitivity where chromatographic resolution is inadequate. This phenomenon has become evident in the case of trace analysis to measure environmental contamination by TCDD. Here, the chemist is looking for low picogram amounts of material in the presence of ubiquitous interfering polychlorinated biphenyls, and both mass and chromatographic resolution must be maximized (5).

Increased scan speed capability is only one of the improvements in the mass spectrometers used in modern GC/MS instruments. Pumping speeds have been increased to handle the gas flows from GC interfaces and from CI operation. Smaller magnets lead to more compact sector instruments, and the engineers have learned to make hyperbolic quadrupole rod assemblies and to minimize fringing field problems. Dual EI/CI sources are· so highly developed that manufacturers boast of switchover times of a few seconds. Data systems have progressed from simple recorders with mass markers through minicomputers with AD/DA converters to microprocessor controlled spectrometers coupled to microprocessor controlled chromatographs all answering to a powerful minicomputer and tied via modem to elaborate central data bases and data reduction systems. The remainder of this chapter will deal less with the fundamental GC/MS advances and more with the enhanced analytical power they offer.

3.   IONIZATION TECHNIQUES

Since 1966, chemical ionization has been a recognized technique of mass spectrometry (6), and its compatibility with GC/MS was immediately obvious (7). The degree of acceptance of CI among GC/MS users is now so great that CI sources are offered by every instrument manufacturer and are often standard rather than optional features. Proton transfer CI from methane, isobutane, and ammonia have been widely used in analytical mass spectrometry, particularly in the biological sciences, where fragile molecules profit from the gentleness of the ionization process.

In environmental analysis, too, proton transfer ionization is popular. The environmental chemist is frequently confronted with an unknown component in a complex chromatogram. To compound the normal difficulty of identifying a substance from its mass spectrum alone, MS sensitivity and GC resolution inadequacies can introduce considerable confusion into that spectrum. In such a case, the independent, complementary information provided by CI can be invaluable (8). The simplicity of proton transfer spectra is also a great help in sorting out the contributions of poorly resolved GC peaks (9). There is some debate whether CI is inherently more sensitive than EI. In terms of recognizing a compound in the presence of interfering species, however, CI is frequently the more sensitive technique because the ions due to both target and interference compounds are limited to a smaller set of masses.

Environmental chemists have accepted proton transfer CI because its simple spectra can facilitate the recognition of a compound in a complex chromatogram, because of its EI-complementary structural information, and because of its ability to selectively ionize or ignore classes of compounds. Even within the category of proton transfer CI, there are variations that provide mutually complementary information. A recent example from environmental analysis is the combination of isobutane-CI with both methane-CI and methane-EI for the unambiguous identification of dicarboxylic acids in photochemical smog (10). A possible application would be in studies of effluent from petroleum processing plants, or of petroleum product contamination where the major portion of the substances present is likely to consist of saturated hydrocarbon, while the more toxic or chemically active species are present in relatively small amounts. An electron impact GC/MS run of such a sample will be dominated by the typical paraffin fragment peaks, and extraction of spectral contributions from more interesting species can be very difficult. If the same sample were run with isobutane-CI, the saturated hydrocarbons would not be ionized, and otherwise minor components would produce the major GC peaks and their spectra would be free of paraffin-caused distortions.

Other forms of selective chemical ionization are being studied in a number of laboratories, and many of them promise to become useful in the complex world of environmental GC/MS. Selective charge exchange from $C_6H_6^+$ to molecules with lower ionization potentials than benzene (9.2 eV) has been reported (11), and the rather more complicated ionization modes of NO^+ reagent may be adopted by environmental analysts as more information on its behavior is amassed (12). Mixed ionization modes are also potentially useful. Hunt (13), for example, combines relatively gentle proton transfer from H_3O^+ with relatively severe charge exchange from Ar^+ to obtain pseudomolecular ions and characteristic fragmentation patterns at the same time. Hites et al. (14) have gone even farther, using the same reagent gas mixture to measure the ratio of charge exchange to proton transfer for polynuclear aromatic hydrocarbons and have been able to distinguish geometric isomers. Such mixed mode CI might find unique applications in revealing the relative amounts of such difficult to separate species as benzo(a)pyrene and benzo(e)pyrene, which are otherwise indistinguishable by mass spectrometry. In general, moving from EI to CI techniques for GC/MS applications has been found to involve compromise of sensitivity or quantitative reproducibility. CI applications in environmental analysis tend to be limited to specialized investigations where some very specific feature is utilized.

A more recent development in CI mass spectrometry, and one with the potential to significantly expand the power of GC/MS in the near future is negative ion chemical ionization. Organic mass spectrometry has traditionally concentrated on positive ions for the simple reason that they are formed in abundance by electron impact, while negative ions are scarce. Under chemical ionization conditions, however, that is not necessarily the case. For each primary positive ion formed in a CI source, at least one low-energy electron is produced. If a species M is present with a high electron-capture cross section, most of these electrons will be converted to negative ions M^-.

Two basic cases are then possible. M could be the "bath gas" which becomes a negative "reagent gas" and plays a role analogous to the familiar positive reagent gasses in ionizing sample via a series of ion–molecule reactions. O_2^-, O^-, Cl^-, C_2H^-, CH_3O^-, $CH_2NO_2^-$, OH^-, and CCl_3^- are among the negative ions that have been used as reagents (15). If the bath gas has a very low electron affinity (e.g., methane or argon), then the thermal electrons themselves are the reagent, ionizing sample molecules by electron attachment. For a complex set of reasons, including the excess of negative charges present, the high mobility of a thermal electron, and the relative rates of recombination, substantially higher currents of negative sample ions than positive sample ions can be produced in a CI source. Hunt (16) using an ion source that allows essentially simultaneous observation of positive and negative ions, has reported negative sample ion currents as much as 10^3 times the positive sample currents for certain compounds. This is illustrated by the data in Table 1 (from Reference 14).

The species to which this sensitivity enhancement applies are often those of greatest interest to the environmental analyst. They include most sulfur- and halogen-containing compounds, metal complexes, and a variety of organic functional groups such as cyanides, carbonyls, acids, and aromatics, in short, the full range of species for which the electron-capture GC detection is advantageous. Risby and Prescott (17) have recently exploited this phenomenon in an environmental analysis context by monitoring ruthenium, platinum, and palladium in automobile exhaust by negative ion GC/MS after producing a volatile complex with a strongly electron-capturing ligand. The thousandfold enhancement in sensitivity toward polychlorodibenzo-p-dioxins of electron attachment CI over both EI and positive methane-CI has been applied to the measurement of several dioxins in animal tissue and wood samples (5). The same authors also investigate the specificity of several negative CI modes for this analysis.

At present, there is only limited information available on the behavior of most substances under negative CI conditions. Study has just begun to reveal the selective ionization capabilities of potentially useful reagent gases (18), the structural information contained in negative CI spectra, and the analytical advantages of combining information from negative CI, positive CI, and EI. This is mainly because major modifications are necessary with existing positive ion instruments, including power supply replacement and, usually, floating of electron multiplier and preamplifier, so little negative ion work has been attempted in applied analytical laboratories. The potential sensitivity gains, however, have been widely noted and have generated so much interest among GC/MS users that most instrument manufacturers now offer negative CI capability in their newer models and as updates to their older instruments. As these instruments get into the hands of more analytical chemists, the potential negative GC/MS analytical power will probably be developed to a level comparable to positive CI.

As the pressure within a CI source is increased from the usual 1 Torr toward 1 atm, the number of collisions a sample molecule undergoes increases proportionately, as does the residence time of a reagent ion (or electron). At atmospheric pressure, every sample molecule will probably collide with a reagent ion, and, if an appropriate exothermic or thermoneutral reaction pathway exists, most of them will be ionized. By contrast,

TABLE 1. Comparison of Positive and Negative Sample Ion Currents Obtained by the PPNICI Technique Using Methane as the CI Reagent Gas

Compound	Molecular Weight	N/P^a	% Total Positive–Negative Sample Ion Current		Other
			$(M)^-$	$(M + 1)^+$	
2,4,6-Trinitrotoluene (TNT)7	227	1042	62.5	0.06	197^- (25.0),210^- (12.5)
O,O,-Diethyl-O-p-nitrophenylthiophosphate (parathion)8	291	102	61.3	0.6	138^- (13.6),154^- (19.1),169^- (3.8)
Pentafluorobenzonitrile 9	193	953	95.3	0.1	—
Pentafluorobenzanilide 10	287	443	3.6	0.1	286^- (29.4),267^- (44.3),266^- (22.7)
N-(pentafluorobenzylidene)aniline 11	271	1021	91.9	0.09	167^- (8.0)
Phenyl pentafluorobenzoate 12	288	682	88.6	0.13	195^+ (0.06)
Benzil 13	210	495	99.0	0.2	105^+ (0.6)
Phthalic anhydride 14	148	124	99.2	0.8	—
9,10-Anthraquinone 15	208	166	99.4	0.6	—

[a] Ratio of total negative sample ion to total positive sample ion current.

only one in 10^3 or 10^4 sample molecules is ionized in conventional CI or EI. This gives atmospheric pressure ionization (API) potentially the ultimate sensitivity attainable by GC/MS. The phenomenal sensitivities of electron-capture GC and of plasma chromatography are due to the same high probability of ionizing collisions.

Several reports of GC/MS with API sources have been published (19), and sources are commercially available. Trace analysis workers have been encouraged by successful API GC/MS detection of femtogram (10^{-15} g) quantities of sample. Nonetheless, the technique has not lived up to its analytical potential. One reason is, incongruously, the very sensitivity of the technique. A trace contaminant in the API–GC/MS system can consume the bulk of the reagent ion supply and even steal charge from the legitimate sample. API is a difficult technique to apply. Therefore, it has not become an established tool of the environmental analysis community despite its successful employment by some of the pioneering groups. API is, however, an active research field with several groups now building and using instruments. Many of the serious drawbacks associated with ion clustering and cluster break-up, energy spread due to extraction across a pressure gradient, and so on, are now understood and are being overcome. The future of API, at least for fairly simple systems, looks bright.

Field ionization mass spectrometry (FIMS) is another gentle ionization technique that minimizes fragmentation. Although GC/FIMS has been reported, it is not very promising as a technique of routine environmental analysis.

High-resolution mass spectrometry offers an exceptionally high specificity of detection because the information obtained can include the elemental composition of the detected ion. GC/MS operation with high-resolution spectrometers is not uncommon (20, 21), and very high sensitivities have also been reported in this mode. Ten picograms of dimethylnitrosamine has been determined (22) by a selected ion monitoring technique where all possibility of interference from other ions is eliminated by the high resolving power of the mass spectrometer. The application to environmental problems is limited, however, by the difficulty of maintaining high-resolution spectrometers when analyzing anything but very clean samples, which are not available in most environmental analyses.

4. DATA HANDLING

If it is only desired to obtain a mass spectrum of one or a few major peaks in a GC run, it is feasible to use a GC/MS instrument without a data system. In environmental analysis, however, highly complex chromatograms are the rule, and if interest is focused on a small number of peaks, they are probably trace components. The mass spectrometer is usually scanned repetitively for most of the GC run at rates up to several spectra per second. To record data at this rate, to store it in conveniently retrievable forms, and to present it for interpretation later, a fairly sophisticated data system is indispensable. Once the data are under the auspices of a data system, the computer can be used to perform a large portion of the reduction, quantitation, and interpretation.

The basic GC/MS data system consists of a dedicated minicomputer with some sort of bulk storage, usually a disk (hard or floppy), and an assortment of input/output (I/O) devices. The interface to the instrument may be just an AD/DA converter, but recent instruments add a microprocessor which is used to control the mass spectrometer and process the data as far as peak identification, peak intensity measurement, and mass assignment. Because the microprocessor assumes so much of the real time work load, the central processing unit is free to process current data (display or even search operations) essentially in real time, or to be used in a foreground–background mode for off-line data reduction or other laboratory purposes.

Following the actual data acquisition and storage, the computer serves several interrelated functions: (1) to organize and present the data in a format that the operator can quickly and conveniently inspect; (2) to enhance the data by manipulating it to minimize distortions and emphasize significant features; (3) to interpret the data by assigning an identity or a partial structure to some or all GC peaks (23); and (4) to use the data in quantitative calculations, comparison with data compilations, and so on.

The quality of the GC/MS data is both a primary stimulus for computer data manipulation and an important factor in limiting the scope of reliable computer interpretation. A high-quality mass spectrum is as much a physical property of a compound as its melting point. If run under standard conditions, the recorded spectrum will always contain the same peaks with the same relative intensities. To obtain physically characteristic spectra, however, one must take pains to ensure that only a single compound is present in the ion source, the scan must be slow enough that statistical fluctuations in ion current are removed by signal averaging, and all operating conditions (e.g., source pressure) must be controlled such that instrumental mass discrimination is absent or constant.

These conditions are seldom obtained in GC/MS analysis of complex samples. "Extra" peaks arise from column bleed or other background and from unresolved components. Peak detection algorithms work best when mass spectrometers are clean and carefully tuned, but the heavy usage of GC/MS instruments too often results in their operation under less than ideal conditions. The resulting distortion of peaks (particularly troublesome with the quadrupole instruments, which represent a large fraction of the environmental GC/MS market) can lead to inaccurate mass assignment and intensity measurement, or even outright rejection by the algorithm, and fast scanning tends to emphasize these distortion problems.

Data systems are used to improve GC/MS spectra that have been distorted by the factors discussed earlier. Constant or slowly varying background is routinely removed by simple subtraction of a spectrum that the operator or the computer decides is representative of background. A measure of *ex post facto* signal averaging can be achieved by summing a set of spectra believed to contain the spectrum of a given component.

Interpretation of GC/MS data usually begins with inspection of a chromatogram reconstructed by the computer from mass spectral data. This provides the operator with immediate access to the chromatographic information concerning sample complexity, success of GC separation, and even concentrations. More limited mass chromatograms can then be used to locate specific compounds or to demonstrate their presence, absence, or relative abundance. For example, the presence of a homologous series of alkyl pyridines in a GC/MS run is strongly suggested if mass chromatograms at $m/e = 93, 107, 121$, and so on, show a regular series of peaks. Alternatively, their absence is immediately demonstrated. Any modern data system with CRT readout can produce the necessary mass chromatograms in a few seconds, or in real time as the data is acquired. Peak integration for quantitative measurement is also best done from mass chromatograms, since the chances for background influence are minimized. This whole sequence can be left to the computer in cases where similar samples are run in large numbers, and only when unexpected peaks occur does the data system call on the operator for more extensive analysis of the data (24).

A prime example of letting the computer do most of the work is in the field of target compound analyses, most familiar as priority pollutant analysis, where many complex samples must be searched for each of a large set of compounds. The process would be prohibitively cumbersome for an operator to carry out by hand. The computer, however, can very rapidly scan through a data file to locate spectra that are similar to each target compound spectrum, measure peak area or height of an appropriate quantitation mass, and even calculate the area relative to an internal standard. The output then lists a relative area for each target compound identified and an index to the reliability of the identification.

Library search routines are the most common mechanism for assigning a greater portion of the interpretive work to the computers, and it is here that data quality becomes a critical limitation. The common form of a search is to compare a given spectrum for the GC/MS run with all spectra, or a preselected subset, in a large library. A statistical index of the similarities of unknown to reference spectrum is then computed, and the most similar reference compounds are presented to the operator as possible identifications of the unknown compounds. The problem arises from the possibility that the unknown spectrum may be less similar to its actual library representation than to other library entries, due to extra peaks, missing peaks, or distorted relative intensities. Alternatively, the unknown may not be represented, but may be similar to library entries. There are also many errors in the commonly used libraries (25), and a glance at the redundant entries in a data compilation will reveal major discrepancies. These range from outright misidentifications to inclusion of large background peaks to varying mass scan limits to such instrumental effects as source temperature. The similarity index reflects only on the two sets of mass/intensity pairs and has no direct way of evaluating spectrum quality. Another index is needed, an index of skepticism based on such outside information as the chemical history of the sample (has an acid wash already removed strong bases?), the GC column properties (could tetrahydrocannibinol be eluted from OV-17 at 120°C?), and simple common sense (why should tobacco smoke contain tetramethyltin?). In specific cases (26), this additional information has been considered in computer data reduction; for instance, retention indices have been combined with mass spectral searches, as has limiting the library searched to a specific molecular weight range or allowed elemental composition. The point of this discussion is to emphasize that computer interpretation of GC/MS data should be approached cautiously and the statistical similarity index should not be accepted as meaningful apart from a critical evaluation of the data by an experienced chemist or spectroscopist.

The proper use of search routines is in a context more limited than an attempt to have the interpretive work taken over by a computer. For example, letting a computer search of a specialized data base pick out chlorinated pesticide residues in a soil extract is a great help in pesticide analysis, providing a reasonable characterization of the sample in a short time while freeing the chemist to concentrate on ambiguous cases and unknown components. The computer can also "reverse search" the data for some given set of masses characteristic of a compound of interest (24, 27). In the case of a completely unfamiliar spectrum, the computer can search through a much larger data base than a human can reasonably handle and condense the possible matches into a smaller set of spectra that the operator can use as a guide for his or her interpretation. Even if the library does not contain a spectrum of the unknown, search results contain useful information. For instance, if the list of the most similar spectra contains several unsaturated fatty acid esters, the operator's attention is directed toward their typical structural features even if fatty acid esters are completely outside his experience. Much more extensive use of the computer's interpretive capabilities is available to the analytical chemist than simple library search routines. The best known of these is the STIRS (28) program, which seeks to extract structural information from mass spectra that do not match any library entries. Such computer techniques can be a tremendous help where the chemist's interpretation capability is limited by time or experience.

A properly designed search routine should feed back to the operator some information on how an unknown spectrum differs from the reference. This can take the form of displaying the two spectra, or their differences, so that the operator can decide on the significance and meaning of the differences, or it can result from an interactive search where the operator first decides what features are significant and directs the computer to search for specific criteria.

More elaborate techniques than simple background subtraction for the computer enhancement of GC/MS data are currently an active field. The best known trends in this direction are the so-called Biller–Biemann dynamic reconstruction procedures (29). They combine the removal of slowly or erratically varying background with what might be called a dynamic mass chromatogram construction. This construction incorporates a decision, based on the data itself, about which masses are significant in a given region of the chromatogram. Basically, such routines look for sets of masses that rise, peak, and decline together in a manner which would be expected from a GC peak. The twofold purpose is to determine which masses are associated, presumably as contributions from a single GC eluant, and to precisely define the elution time of that component even when the total ion chromatogram shows no distinct peaks (the latter is also a property of a mass chromatogram). The first generation of these enhancement routines had limited success in producing clean spectra, suitable for library search, out of messy GC/MS chromatograms, but the routines have proven very useful in locating GC peaks in those chromatograms and especially in indicating the degree of overlap, thus showing the operator where to expect problems. The computer enhancement of GC/MS data is receiving considerable attention and several routines are now available to GC/MS users from data system vendors or other sources (30). Successful use of these procedures to prepare otherwise

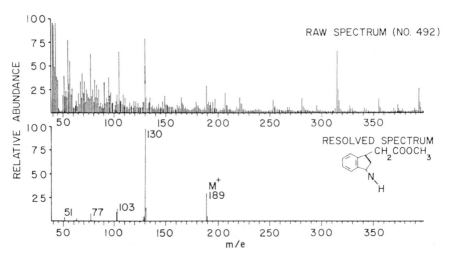

Figure 2.   Example of computer data enhancement to extract a useful spectrum from badly contaminated raw data. Reprinted with permission from R. G. Ramsey et al., *Anal. Chem.* **48,** 1368 (1976). Copyright 1976 American Chemical Society.

unsatisfactory data for interpretation by a cautious chemist or even for computer search is becoming common and should prove a boon to the environmental chemist who could use some help in his habitual search for needles in haystacks. A dramatically successful example of data enhancement is shown in Figure 2 (from Reference 29), where a clean, useful spectrum is extracted from a hopelessly contaminated raw spectrum.

5. ALTERNATIVE INSTRUMENT CONFIGURATIONS

The most common means of introducing a sample into a GC/MS instrument is by syringe injection of a solution containing the species to be determined. The solution is usually obtained by a direct extraction of the environmental sample (a solid–liquid or liquid–liquid partition) as from a tissue homogenate, a sediment or filter-collected particles; or it results from extracting species that have been concentrated on the surfaces of an adsorption or condensation trap.

Only where inherently volatile samples such as head space adsorbates are concerned are we dealing with a sample that is immediately and entirely compatible with mass spectrometry. It has been established, in fact, that only about 10% of the organic species in a typical water extract are sufficiently volatile to pass through a GC column (31). The other 90%—humic acids, lipids and other high molecular weight or extremely polar materials—are generally not suitable for mass spectrometric analysis and should be excluded from the ion source, which they would quickly foul. Since the gas chromatographic column is an effective filter for the removal of such undesirable materials, however, these raw extracts can often be run on a GC/MS without a separate cleanup step, as long as the operator is alert for injection port pyrolysis products. GC injection port and column cleaning are much easier and cheaper than ion source rejuvenation.

An alternative means of sample introduction is the thermal desorption of volatiles from the adsorbent material directly in the GC inlet. This is usually done by rapidly heating the sorbent to 250°C or so while the column is maintained at a low temperature, so that all desorbed species collect as a slug at the column head, and a normal GC run ensues. Equipment for doing this is commercially available, and facilities for the adsorption (sampling) step are included.

With an extension of this technique to higher temperatures, information on the less volatile portion can also be obtained. Direct pyrolysis GC/MS characterization of involatile samples is a means of bypassing the extraction and concentration steps. Since this means following the characteristic destruction of a sample rather than selecting a portion of that sample for analysis, the information obtained is not necessarily equivalent, but it is no less meaningful for many purposes (32). Coal, polymers, and even bacteria are characterized by pyrolysis GC/MS, and there is no reason why sediments, sludges, collected aerosols, and other primary environmental samples can't be handled in the same way. The pyrograms of such materials are just as characteristic as their extract chromatograms.

GC/MS characterization of a complex sample need not involve identification of all major components. One can take advantage of the selective detection powers of the mass spectrometer to produce chromatograms characteristic of its interesting features. Mass chromatograms reconstructed from a scanning GC/MS run are one way to do it. Selected ion chromatograms, where only a small set of masses is monitored, are another, with the added advantage of much increased sensitivity. Alternatively, selective modes of ionization can be used to emphasize interesting components at the expense of others. Proton transfer from relatively weak acid reagent gases such as isobutane and ammonia have been used, and so have relatively weak charge exchange reactants, including NO^+ and $C_6H_6^+$.

Another separation technique that has become popular in environmental analysis is HPLC. In principle, the combination of liquid chromatography (LC) with MS should provide analytical power comparable to what GC/MS has achieved. Unfortunately, LC and MS are much less compatible than are GC and MS, not only in operating conditions but in the nature of the substances for which they are appropriate. The only major incompatibility of GC and MS was in operating pressure, and this has been essentially completely overcome by means of molecular separator and vacuum system design. The "carrier" flow in a liquid chromatograph is usually greater than 0.5 cm^3/min of liquid, which corresponds to at least 300 cm^3/min of gas, or about 10 times the normal GC carrier flow. Furthermore, the carrier is methanol, water, hexane, or some other liquid rather than helium so the molecular separators used in GC/MS will not work efficiently even if the eluant can be vaporized.

Two types of LC/MS interfaces have been thoroughly developed to date. The first is the result of the deposition of eluant on a moving wire or ribbon (33) that transports it through a vaporization chamber where solvent is boiled off, then to the spectrometer ion source where sample evaporates and is ionized, as from a normal solids probe. The second results from splitting (34) the eluant and directing a manageable fraction of it (1–10%) through a heated capillary into the CI ion source, using LC solvent as reagent gas. An interesting variation on the latter approach (35) is to use a "micro" LC at a flow rate of 0.008 cm^3/min with the entire eluant run through the normal jet separator, again using solvent CI.

Both these approaches are reasonably straightforward and, if one is willing to accept the obvious limitations on, for instance, sensitivity and, as CI reagent selection, both are acceptable means of LC/MS coupling for routine analytical purposes. The greatest limitation, however, is inherent in the vaporization of eluant, both solvent and solute, by contact with heated surfaces. This eliminates what is probably the primary reason for using LC as a separation technique: LC works on species that are not sufficiently volatile for GC analysis or that decompose on the hot surfaces of a GC. Only if the limitations inherent in vaporizing labile or involatile molecules can be overcome by these types of interface is LC/MS likely to become an analytical technique that goes significantly beyond what can be done by GC/MS. Such limitations are, of course, long recognized problems of mass spectrometry and the subject of much research. Perhaps the use of rapid heating, fast atom bombardment, field desorption, desorption from special surfaces, or surface ionization will someday make the mass spectrometer as versatile a detector for LC as it now is for GC. Other means of LC/MS coupling are under development, for instance using a laser to vaporize eluant without surface contact. It remains to be seen whether routinely useful analytical tools can be created by such means.

By their analogy to GC/MS, other separation–MS analytical schemes deserve mention here. One of these is controlled distillation from a slowly heated probe in a mass spectrometer ion source. This is by no means a new technique, but has recently received renewed interest as CI and computer data handling facilities become more prevalent. Risby and Yergey (36) have reviewed the technique where pyrolysis rather than simple distillation occurs. The technique may be useful in any case where pyrolysis–

GC/MS is appropriate, but the difficulty of extracting information from the direct pyrolysis data will probably preclude any general application except as a fingerprinting technique. A rapidly developing field is the so-called MS/MS technique. A primary mass analyzer separates ions of selected mass from a complex mixture spectrum and directs the selected ion beam into a collision region where the ions are further fragmented. A second mass analyzer is used to obtain a mass spectrum of the selected primary ion. MS/MS instruments are commercially available. Their greatest promise in environmental applications would seem to be as a screening technique to determine whether a specific compound is present in a mixture. The great advantage is that there is no need for extensive sample cleanup. A wealth of other information is potentially available from MS/MS, and future developments may well turn it into one of the more commonly used analytical tools.

The past ten years have witnessed major development in the instrumentation and methodology of GC/MS. Whether this pace will continue is an open question, but that the results will continue to benefit the environmental chemist is a certainty. More and more environmental analysts are obtaining access to the enhanced capabilities of GC/MS, and it is in the application of these to their specific environmental problems that the significant developments of the next five years are to be expected.

REFERENCES

1. R. J. Leibrand, *J. Chromatogr. Sci.* **13**, 556 (1975).
2. G. M. Janini, G. M. Muschik, J. A. Schroer, and W. L. Zielinski, Jr., *Anal. Chem.* **48**, 1879 (1976).
3. M. Novotny, "Capillary Gas Chromatography in the Analysis of the Environment," in *Environmental Aspects of Analytical Chemistry*, D. F. S. Natusch and P. K. Hople, Eds., Wiley, New York, 1982.
4. R. M. Bean, P. W. Ryan, and R. G. Riley, in *High Resolution Chromatography*, S. Cram, Ed., Academic, New York, 1978.
5. J. R. Hass, M. D. Friesen, D. J. Harvan, and C. E. Parker, *Anal. Chem.* **509**, 1474 (1978).
6. F. H. Field, *Acc. Chem. Res.* **1**, 42 (1968).
7. D. M. Shoengold and B. Munson, *Anal. Chem.* **42**, 1811 (1970).
8. L. S. Sheldon and R. S. Hites, *Environ. Sci. Technol.* **12**, 1188 (1978).
9. F. L. Schulting and E. R. J. Wils, *Anal. Chem.* **49**, 2365 (1977).

10. D. Grasjean, K. Van Cauwenberghe, S. P. Schmid, P. E. Kelley, and J. N. Pitts, Jr., *Environ. Sci. Technol.* **12**, 313 (1978).

11. S. C. S. Rao and C. Fenselau, *Anal. Chem.* **50**, 511 (1978).

12. D. F. Hunt and J. F. Ryan III, *J. Chem. Soc. Chem. Commun.*, 620 (1972).

13. D. F. Hunt and J. F. Ryan III, *Anal. Chem.* **44**, 1306 (1972).

14. M. L. Lee and R. A. Hites, *J. Am. Chem. Soc.* **99**, 2008 (1977).

15. K. R. Jennings, in *Mass Spectrometry*, Vol. 4, R. A. W. Johnstone, Ed., The Chemical Society, London, 1977, p. 203.

16. D. F. Hunt, G. C. Stafford, Jr., F. W. Crow, and J. W. Russell, *Anal. Chem.* **48**, 2098 (1976).

17. S. R. Prescott and R. H. Risby, *Anal. Chem.* **50**, 562 (1978).

18. K. R. Jennings, in *High Performance Mass Spectrometry*, ACS Symposium Series 70, M. L. Gross, Ed., American Chemical Society, Washington, D.C., 1978, p. 179.

19. I. Dzidic, D. I. Carroll, R. N. Stillwell, and E. C. Horning, *Anal. Chem.* **48**, 2098 (1976).

20. K. P. Evans, A. Mathias, N. Mellor, R. Sylvester, and A. E. Williams, *Anal. Chem.* **47**, 821 (1975).

21. E. J. Gallegos, *Anal. Chem.* **47**, 1150 (1975).

22. K. R. Compson, S. Evans, D. Hazelby, and L. E. Moore, in *Proceedings of the Twenty-Fifth Annual Conference on Mass Spectrometry and Allied Topics*, 1977, p. 178.

23. F. W. McLafferty, *Anal. Chem.* **49**, 1440 (1977).

24. S. C. Gates, M. J. Smisko, C. L. Ashendi, J. F. Holland, N. D. Young, and C. C. Sweeley, *Anal. Chem.* **50**, 433 (1978).

25. D. D. Spark, R. Venkataraghavan, and F. W. McLafferty, *Org. Mass. Spectrom.* **13**, 209 (1978).

26. R. A. Strauss and R. H. Hartel, *J. Chromatogr.* **134**, 39 (1977).

27. D. W. Kuehl, *Anal. Chem.* **49**, 521 (1977).

28. K. S. Kwok, R. Venkataraghavan, and F. W. McLafferty, *J. Am. Chem. Soc.* **95**, 4185 (1973).

29. J. E. Biller and K. Biemann, *Anal. Lett.* **7**, 515 (1974).

30. R. G. Ramsey, M. J. Stefik, T. C. Rindfleisch, and A. M. Duffield, *Anal. Chem.* **48**, 1368 (1976).

31. L. H. Keith, in *Identification and Analysis of Organic Pollutants in Water*, L. H. Keith, Ed., Ann Arbor Science Publishers, Ann Arbor, 1976, p. iv.

32. T. T. Coburn, R. E. Bozak, J. E. Clarkson, and J. H. Campbell, *Anal. Chem.* **50**, 958 (1978).

33. W. H. McFadden, H. L. Schwartz, and S. Evans, *J. Chromatogr.* **122**, 389 (1976).

34. F. W. McLafferty, R. Knutti, R. Venkataraghavan, P. J. Arpino, and B. G. Dawkins, *Anal. Chem.* **47,** 1503 (1975).
35. T. Takeuchi, Y. Hirata, and Y. Okumura, *Anal. Chem.* **50,** 659 (1978).
36. T. H. Risby and A. L. Yergey, *Anal. Chem.* **50,** 326A (1978).
37. F. A. Thome and G. W. Young, *Anal. Chem.* **48,** 1423 (1976).

4

ENVIRONMENTAL APPLICATIONS OF SURFACE ANALYSIS TECHNIQUES: PAS, XPS, AUGER ELECTRON SPECTROSCOPY, SIMS

R. W. LINTON, D. T. HARVEY, and G. E. CABANISS

Department of Chemistry
The University of North Carolina
Chapel Hill, North Carolina 27514

1. INTRODUCTION

The coupling of various microprobe excitation sources (e.g. laser, ion, or electron beams) with analytical spectrometers capable of spatially resolved chemical measurements represents a recent evolutionary, if not revolutionary, trend in analytical chemistry. The sudden proliferation of analytical techniques capable of *in situ* microscopic measurements is having an extraordinary impact on the applied sciences, including environmental chemistry, which serves as the specific focus of this review.

The ideal *in situ* measurement would provide sensitive quantitative characterization of elements and compounds, including their physical distribution both in the surface plane of a solid sample and as a function of depth. The ideal technique also would not be subject to analysis artifacts introduced by sample preparation and presentation to the spectrometer or by the inherent fundamental physical limitations of the spectroscopic technique. The substantial array of current microanalytical and/or surface instrumentation attests to the lack of superiority of any given technique with respect to *all* of the preceding criteria.

Various surface analysis techniques have been applied to problems of environmental interest. Emphasis in this review is given to surface techniques capable of particulate characterization that provide either (1) chemical speciation information about environmental contaminants or (2) multielemental composition profiles of individual environmental particles with high three-dimensional spatial resolution. The former category will be represented by x-ray photoelectron spectroscopy (XPS) and photoacoustic spectroscopy (PAS). The latter will be discussed in the context of Auger and ion microprobe techniques that are microanalytical versions of Auger electron spectroscopy and secondary ion mass spectrometry (SIMS), respectively. Other important microanalytical spectroscopies, including laser, Raman, and electron microscopic or microprobe techniques, will not be discussed because of availability of other recent reviews including environmental applications (1–3). Furthermore, the conventional electron and laser microprobe-based techniques generally are not capable of providing *surface* analyses with depth resolutions approaching that of XPS, Auger electron spectroscopy, or SIMS measurements.

The major environmental application of surface analysis techniques has been in the investigation of particulate materials and the interfacial chemistry governing the partitioning of particle-associated toxic species between various environmental compartments. For example, the binding of toxic trace metal species by waterborne sediments is of interest with respect to environmental transport and bioavailability. Similarly, airborne particles serve as a site for the condensation or surface adsorption of volatile materials, including those generated in high temperature industrial or combustion sources. Examples of surface-associated materials are nitrogen and sulfur oxides, volatile trace metals, and sorbed organic species such as polynuclear aromatic hydrocarbons (PAH). The particle surface may serve, therefore, as a specific site for heterogeneous catalytic reactions, including atmospheric acid nitrate and sulfate salt formation, and for photochemical reactions, including the photodecomposition or chemical transformation of sorbed PAH. The potentially toxic surface species are directly available to the environment via atmospheric washout; groundwater leaching; or extraction by biological fluids, cells, and tissues. Pollution control technologies also are least effective in controlling the smaller size fractions (<3-μm diameters) of airborne particles, which usually have the highest surface area to volume ratios. Such particles have considerable atmospheric residence times, are in the respirable size range, and tend to be most enriched with respect to the average concentrations of toxic species as the consequence of the surface deposition processes and the size-dependent surface area to volume ratios just cited.

In summary, knowledge of the surface chemistry of environmental particles is potentially useful in the study of:

1. Mechanisms of particle formation and environmental dispersion.
2. Chemical transformations of particles in various environmental compartments.
3. Physicochemical parameters governing the bioavailability of toxic species.
4. Improved particle emission control technologies.

This chapter will review the utility of existing surface analytical techniques primarily in the context of the aforementioned aspects of environmental chemistry.

2. PHYSICAL BASIS OF TECHNIQUES/INSTRUMENTATION

2.1. Photoacoustic Spectroscopy (PAS)

Although the photoacoustic effect was discovered over 100 years ago (4), it was of little importance until advances in microphone technology were made. Gas phase infrared photoacoustic determinations of CO and CO_2 were successful as early as the late 1930s (5, 6). Even with the somewhat inefficient detectors of that time, sensitivity was sufficient to detect 0.2% by volume CO_2. Condensed phase PAS was "rediscovered" by Parker (7) as an interference in gas phase experiments (in the form of weak signals from cell windows) (8).

The photoacoustic effect is induced by nonradiative excited state relaxation processes. Heat is produced. This heat may diffuse through a liquid or solid until it reaches a gaseous interface where it is transferred to the gas in the form of kinetic energy. If the excitation source is chopped or repetitively pulsed, the result is modulation of the heat flow to the gas-condensed phase interface. The heat exchange along this interface causes pressure changes in the boundary layer of gas, making it act like a small piston acting on the bulk of the gas (9, 10). If the sample and surrounding gas are enclosed in an airtight cell, this piston action may be detected either by a sensitive microphone or a piezoelectric detector in contact with the condensed phase. The PAS technique is particularly useful in studies of optically opaque or highly light scattering materials.

Photoacoustic instrumentation has been used in the ultraviolet–visible (UV–VIS), near infrared (near IR), infrared, and microwave regions of the electromagnetic radiation spectrum. Complete commercial UV–VIS–near IR instruments were marketed in the late 1970s (e.g., EG&G Princeton Applied Research Model 6001 and Gilford Instruments Model R1500). Princeton Applied Research currently supplies the basic components for construction of a photoacoustic system.

A block diagram of a typical UV–VIS–near IR PAS system is shown in Figure 1 (11). In this UV–VIS system, source compensation is accomplished by ratioing the output from the photoacoustic cell to that of a pyroelectric detector in a parallel channel. This design could be modified for true double beam operation by replacing the pyroelectric detector with another photoacoustic cell. Alternatively, pseudo–double beam operation

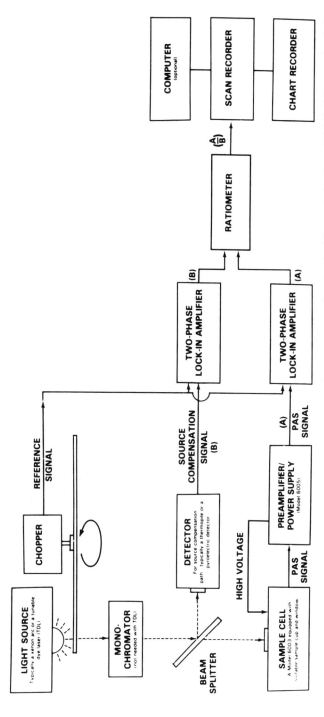

Figure 1. Block diagram of a commercial UV-VIS photoacoustic spectrometer. Reprinted with permission from EG&G, Princeton Applied Research (11).

141

may be accomplished by storing a reference (blackbody) spectrum in computer memory for utilization in source compensation routines.

Although dispersive infrared photoacoustic spectroscopy has been reported (12–15), the generally lower absorptivity of vibrational modes in IR relative to electronic transitions in UV–VIS limits the sensitivity of IR–PAS. This limitation has been partially circumvented in IR–PAS by use of extremely intense carbon rod furnace sources (16). However, other authors (17–27) have successfully demonstrated the Fellgett (multiplex) and Jaquinot (throughput) advantages available from Fourier transform infrared photoacoustic spectroscopy (FT–IR–PAS).

There are two commercially available cells for use in FT–IR–PAS. The output signal from the FT–IR–PAS detector is directed to the analog-to-digital converter of a standard fast scanning FT–IR spectrometer. A block diagram of the Princeton Applied Research system including a model 6003 sample cell and a model 6005 photoacoustic detector is shown in Figure 2. Infrared source radiation is modulated by the constructive and destructive interference that occurs as the light returning from the fixed and movable mirrors recombines at the beam splitter of the interferometer. This modulation is in the audiofrequency range and varies with wave number by the following relationship (28):

$$f = 2V\bar{v} \tag{1}$$

where f is modulation frequency (sec^{-1}), V is mirror velocity (cm sec^{-1}),

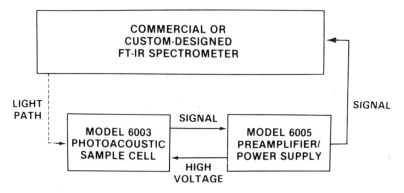

Figure 2. Block diagram of an FT-IR photoacoustic spectrometer. Reprinted with permission from EG&G, Princeton Applied Research (11).

and $\bar{v}$ is the wave number (cm^{-1}). Typical f values [Digilab FTS-15 (23)] range from 240–2400 Hz for 400–4000 cm^{-1}. Therefore, FT–IR–PAS interferograms produced by sources modulated by fast scanning Michelson interferometers may, after preamplification, be processed with existing hardware and software of conventional FT–IR spectrometers. Source compensation in FT–IR–PAS may be achieved by deconvolution involving division of the spectrum by the source envelope in the frequency domain. The source envelope may be obtained by acquiring a spectrum of a blackbody absorber. Carbon black has been used, but adsorbed contaminants (H_2O, CO_2) make this technique somewhat unreliable (15, 24). Alternatively, the source envelope may be obtained by acquiring the response of the triglycine sulfate (TGS) detector under normal FT–IR operation. However, this method does not account for the frequency response of the cell (23).

Fourier transform UV–VIS PAS has been investigated (29). Conventional UV–VIS absorption spectroscopy generally does not benefit from Fourier transform techniques since source stability is the limiting factor with respect to noise (loss of Fellgett advantage). However, FT–UV–VIS PAS is usually detector noise limited. Therefore, throughput and multiplex advantages may be realized.

2.2. X-Ray Photoelectron Spectroscopy (XPS)

In x-ray photoelectron spectroscopy (XPS) a nearly monoenergetic x-ray beam is used to induce the ejection of core-level electrons. Although the energy of the x-ray beam is high enough to penetrate into the bulk of the sample, the escape depth of the photoelectrons is limited by their inelastic mean free paths. The analytical sampling depth typically is less than 40 Å for metals and metal oxides, thus making XPS a surface-specific technique.

The measured kinetic energy (E_k) of an ejected photoelectron can be related to the binding energy (E_b) of the electron to the nucleus:

$$E_b = E_0 - E_k - \phi_w \qquad (2)$$

where E_0 is the initial x-ray beam energy and ϕ_w is the spectrometer work function. Since the binding energy of a core-level electron is sensitive to the electron density in the valence shell, the binding energy shifts in a

XPS spectrum can be used to obtain chemical speciation information about the analyte.

The basic elements of a typical x-ray photoelectron spectrometer are shown in Figure 3. The x-ray source is usually a Mg (1253.6-eV) or Al (1486.6-eV) anode, although higher-energy anodes have been used. X-ray monochromators may be utilized to narrow x-ray linewidths with consequent improvements in the energy resolution of photoelectrons. Ejected photoelectrons may be energy analyzed with electrostatic analyzers of various designs and detected with electron multipliers. As is true of any surface technique involving charged particles, the spectrometer is maintained under an ultrahigh vacuum ($<10^{-8}$ Torr) to ensure that the photoelectrons maintain a long mean free path in the analytical chamber and to minimize surface contamination from residual gases in the spectrometer. More detailed discussions of XPS theory and instrumentation can be found in several excellent references (30–32), including a detailed discussion by Novakov in this text (33).

2.3. Auger Electron Spectroscopy and Secondary Ion Mass Spectroscopy (SIMS)—Auger and Ion Microprobe Analysis

Auger electron and x-ray emission are competitive processes resulting from either x-ray or electron bombardment induced core-level ionization of sample atoms (Figure 4) (34–37). The scanning Auger microprobe and electron microprobe employ focused primary electron beams to achieve lateral resolutions on the order of 1 μm or less. Since Auger electron emission predominates over x-ray emission for low atomic number elements, the Auger electron spectroscopy technique is more sensitive for light elements (Li → Na). For heavier elements, both techniques provide detection limits on the order of 0.1% atomic concentration within the analytical volume. The analytical sampling depths, however, are quite different for the electron and Auger microprobe techniques. Core vacancies resulting from electron bombardment are generated at depths ranging up to micrometers into the sample. The resultant x-ray emission traverses this depth range resulting in rather poor depth resolution for the electron microprobe technique. Analogous to XPS, the escape depth of the Auger electrons is limited by their inelastic mean free paths resulting in highly surface-specific measurements.

A conventional multitechnique instrument capable of ultrahigh vacuum

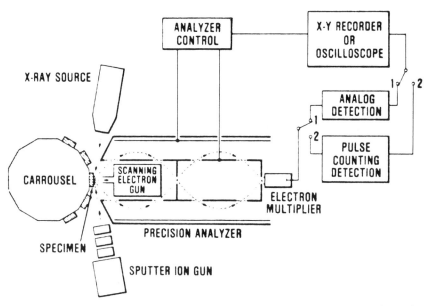

Figure 3. Block diagram of a commercial x-ray photoelectron spectrometer and scanning Auger microprobe (Perkin Elmer-Physical Electronics Industries).

XPS, scanning Auger microanalysis, and ion beam sputter depth profiling is shown in Figure 3. The precision electron energy analyzer is a double pass cylindrical mirror design (CMA) providing the high transmission (~7%) required for good sensitivity in both XPS and Auger electron spectroscopy modes (38, 39). For Auger electron spectroscopy analysis, the

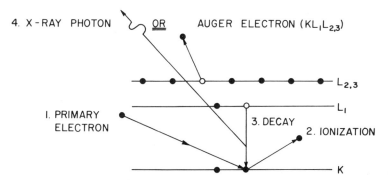

Figure 4. Schematic showing electron-induced Auger electron emission ($KL_1L_{2,3}$ transition shown) and the competitive process of x-ray emission.

analyzer is usually operated in a nonretarding, analog detection mode providing constant energy resolution ($\Delta E/E$) throughout the spectrum. Circular apertures at the outputs of the CMA stages can be switched to vary energy resolution or sensitivity. A scanning electron gun mounted coaxially in the middle of the CMA is utilized for scanning Auger microanalysis. The focused primary beam can be scanned synchronously with an oscilloscope display device to obtain absorbed current or backscattered electron micrographs, Auger images or maps of selected elements, or Auger line scans. The electron gun utilizes electrostatic deflection to eliminate stray magnetic field effects in the CMA.

In SIMS, a beam of primary ions is used to eject surface atoms from a solid sample (40–43). The primary ion–surface interaction may be conceptualized as a series of hard sphere collisions or a collision cascade resulting in momentum transfer and ultimate sputtering of species present within a few atomic layers of the surface. Although only a small fraction of the sputtered species are ionized (often <1%), the ions can be detected using mass spectrometry to provide a sensitive surface analysis. Elemental detection limits range from about 10^{-2}–10^{-7} at. %, depending on the primary ion beam characteristics (energy, current density, composition), the element and sample matrix of interest (sputtering and secondary ion yields), and instrumental considerations (mass spectrometer transmission, energy resolution, mass resolution, etc.).

There are two instrumental configurations capable of performing SIMS measurements with high spatial resolution (41, 44–46). The ion microanalyzer (IMA) (see Figure 5a) and the ion microprobe (IMP) (see Figure 5b). The IMP achieves high lateral resolution by using a focused primary ion beam that currently is limited to diameters ≥ 1 μm due to sensitivity and ion optical considerations. A suitable primary ion condenser lens system should permit useful probe diameters approaching 0.1 μm (47). Ion images are obtained in the IMP in a manner comparable to that used in scanning Auger microscopy. The primary beam is rastered and the resulting secondary ion intensity modulates the intensity of an oscilloscope whose x-y deflection is synchronized with the primary beam raster scan. The IMA produces a magnified ion image via secondary optics that transmit secondary ions to an imaging detector such as a channelplate (46). A one-to-one correspondence is maintained between the lateral distribution of elements on the sample surface and that of the final mass-

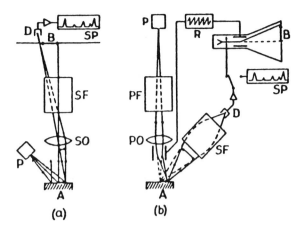

Figure 5. Block diagrams showing two microanalytical versions of SIMS (44): (*a*) Ion microanalyzer—direct imaging combined with analysis. (*b*) Ion microprobe. (A = sampled area; B = image of A; P = primary ion source; SO = secondary optics; SF = secondary filter; D = detector; SP = spectrum; PF = primary filter; PO = primary optics; R = raster scanning.)

resolved ion image. The best secondary optics currently available permit a lateral resolution of 0.3 μm on ideal samples (i.e., high sensitivity element, low mass resolution, minimal surface roughness, no electrical charging of the specimen).

State-of-the-art IMA or IMP instruments incorporate many advanced features required for the optimization of SIMS measurements on chemically complex environmental microparticulates including (1) double focusing mass spectrometers for high mass resolution and accurate isotope ratio measurements, (2) rapid magnetic sector peak switching under microcomputer control for multielement depth profiles on the same sputtered area (e.g., a single environmental particle), (3) digital ion image acquisition and processing, (4) positive and negative ion mass spectrometry, and (5) variable primary beam composition to maximize secondary ion yields for most elements (41–48).

More detailed reviews of the physical basis and instrumentation employed in Auger electron spectroscopy and SIMS are available elsewhere and are beyond the scope of this review of environmental applications (34–48).

3.   REVIEW OF ENVIRONMENTAL APPLICATIONS—PARTICLE CHARACTERIZATION

3.1.   PAS

Although a potentially important applications area for PAS, the study of environmental particles or experimental models of such systems is still very limited. Killinger et al. (29, 49) have used the photoacoustic effect to measure light absorption by airborne carbonaceous particulates (soot). The success of this technique is based on the following: (1) optical absorption per particle mass unit is independent of particle size for spherical particles with diameters much less than the analytical wavelength (50, 51); (2) photoacoustic spectroscopy is not as susceptible to light scattering as optical absorption or transmission techniques; (3) the sensitivity of conventional absorption measurements for the determination of carbonaceous particulate matter is not as high as that of photoacoustic measurements (49).

A block diagram of the device used by Killinger et al. for the comparison of absorption, scattering, and photoacoustic measurements is shown in Figure 6 (49). Absorption and photoacoustic spectra of soot generated by propane combustion were shown to be identical in the 590–620 nm range. The absorption coefficient ($2.1 \times 10^{-3} \text{ cm}^{-1}$) and specific

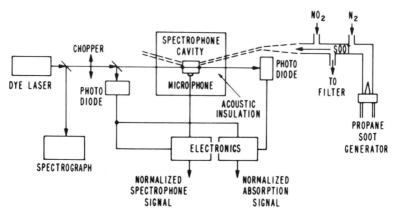

Figure 6.   Block diagram of a spectrometer that measures absorption, light scattering, and photoacoustic response of automobile exhaust or soot (49). Reprinted with permission from K. K. Killinger and S. M. Jaspar, *Chem. Phys. Lett.* **66**, 207 (1979).

absorption coefficient (15 L/g·cm) were calculated using absorption data and standard concentrations. The average soot concentration produced by a diesel engine was shown to be about 50 times higher than a gasoline engine using the photoacoustic detection system.

Other PAS research relating to airborne environmental particulates was recently published by Cabaniss and Linton (51). Polynuclear aromatic hydrocarbons were sorbed on particulates using a fluidized bed adsorption device similar to one proposed by Miguel et al. (52). These particles were used as a model of organic vapor adsorption by combustion generated aerosols produced in sources such as coal-fired power plants. This research will be discussed in more detail throughout Section 4 of this review.

Howell and Palmer (53) have investigated the utility of PAS as a nondestructive tool for the characterization of materials trapped by atmospheric aerosol filters. Carbonaceous species were observed by using UV–PAS, while ammonium sulfate and water were determined by use of vibrational overtones occurring in the near IR region.

Preliminary studies relating to asbestos (chrysotile) determinations in environmental samples have been reported by Monchalin et al. (54). A continuous wave HF laser was used to observe the OH doublet absorbance between 3400–3800 cm^{-1}. Photoacoustic results show that this doublet, characteristic of chemically bound hydroxyl groups, is easily distinguished from the band due to physisorbed water. Another result is that light absorption by these chemically bound hydroxyl groups is dependent on the orientation of the light polarization relative to the asbestos fiber axis.

3.2. XPS (Aquatic Particles)

Since the surface of an environmental particle represents only a small fraction of the particle's bulk, the utility of conventional bulk techniques (e.g., x-ray fluorescence, x-ray diffraction, and Mössbauer spectroscopy) for studying surface interactions is limited. The surface specificity of XPS, coupled with its ability to provide chemical speciation information, makes XPS a promising technique for both atmospheric studies, as reviewed in this text by Novakov (33), and aquatic environmental particulate systems to be discussed herein.

It is well known that suspended matter, in the form of clay minerals (55), metal ion precipitates (55), and hydrous ferric and manganese oxides

(55, 56), can serve as a sink for trace metals and thus play an important role in the transport of metals through aquatic environmental systems. XPS has proven useful in the analysis of both aquatic particles and adsorbed cationic and anionic species.

Koppelman and Dillard (57) have used XPS in conjunction with Mössbauer spectroscopy to determine the bonding nature of iron in the clay minerals chlorite, illite, kalonite, and nontronite. They reported that binding energy shifts for the $Fe(2p_{3/2})$ peak allowed the differentiation of Fe^{2+} in chlorite and Fe^{3+} in nontronite. The XPS quantitative analysis of Fe^{2+} and Fe^{3+} in illite (71% Fe^{3+}), determined by curve fitting, was in fair agreement with Mössbauer results (~80% Fe^{3+}).

The surface of hydrous ferric oxide (HFO) has been characterized with XPS by comparison with several iron oxide mineral standards (58). By applying a curve-fitting routine to the oxygen $1s$ peak, it was possible to distinguish between oxide, hydroxide, and physically adsorbed water oxygens. The hydroxide/oxide ratio for the amorphous fresh HFO sample, determined by XPS, was similar to values determined for other iron oxide minerals known to have a goethite (α-FeOOH) structure (Table 1), suggesting that fresh HFO has a FeOOH structure as proposed by Stumm and Morgan (59) and Kabayashi and Uda (60).

The adsorption of Pb^{2+} on montmorillonite has been studied at pH 6 (61). Comparison of the $Pb(4f_{7/2})$ and $(4f_{5/2})$ peaks with known standards (Pb metal, PbO, PbO_2) suggested that the lead was adsorbed as PbO.

Adams and Evans (62) have studied the cation-exchange capacity of the layered silicate beidellite and investigated the importance of surface versus interlamellar uptake of Na^+, K^+, Ca^{2+}, Pb^{2+}, and Ba^{2+}. They found that the XPS-determined Na^+ and Ca^{2+} uptake were in good agree-

TABLE 1.   XPS and X-ray Powder Diffraction
(XRD) Results for Various Iron Oxides[a]

Sample	OH^-/O^{2-}	XRD Results
Goethite	1.06 ± 0.13	α-FeOOH
Limonite	1.00 ± 0.10	α-FeOOH
Aged HFO[b]	1.10 ± 0.20	α-FeOOH
Fresh HFO	1.17 ± 0.11	amorphous

[a] Data from Reference 58.

[b] Aged for 1 year, HFO = hydrous ferric oxide.

TABLE 2. Binding Energies for Fe($2p_{3/2}$) in Clay Minerals[a]

	Fe^{2+}, lattice (eV)	Fe^{3+}, lattice (eV)	Fe^{3+}, adsorbed (eV)	Δ (eV)[b]
Chlorite	710.3[c]	—	711.4[c]	−1.1
Illite	710.4[c]	712.6[c]	711.5[c]	−1.0
Kalonite	—	—	711.4	−1.1
Nontronite	—	712.5	—	—

[a] Data from Reference 57.
[b] Δ = Fe^{3+}, adsorbed − Fe^{3+}, lattice (Nontronite).
[c] Determined by curve fitting.

ment with bulk values determined by nephelometry, suggesting equal surface and interlamellar exchange. The XPS-determined uptake of K^+, Pb^{2+}, and Ba^{2+}, however, were higher than bulk values, suggesting that, in addition to interlamellar exchange, these cations are strongly held on the edges or surface of the beidellite. When Ca^{2+} and Ba^{2+} were exchanged simultaneously, the XPS results suggested that Ca^{2+} is exchanged in the same manner as the pure Ca^{2+}-exchanged sample, while the Ba^{2+} exchanged only at the surface sites.

Koppelman and Dillard have used XPS to study the adsorption of a variety of metals on the clay minerals chlorite, kalonite, and illite. In a study of Fe^{3+} and Cr^{3+} adsorption (57), they found that the binding energies for the adsorbed ions were substantially lower than the binding energies for the ions in a lattice environment (Tables 2 and 3). They suggest that this is evidence for specific adsorption to negatively charged binding sites on the clay surface. Curve fitting was necessary to determine the

TABLE 3. Binding Energies for Cr($2p_{3/2}$) in Clay Minerals[a]

	Cr^{3+}, lattice (eV)	Cr^{3+}, adsorbed (eV)	Δ (eV)[b]
Chlorite	—	577.2	−1.1
Illite	—	577.3	−1.0
Kalonite	—	577.5	−0.8
Uvarovite[c]	578.3	—	—

[a] Data from Reference 57.
[b] Δ = Cr^{3+}, adsorbed − Cr^{3+}, lattice (Uvarovite).
[c] $Ca_3Cr_2(SiO_4)_3$.

contribution of lattice Fe^{2+} and Fe^{3+}, and adsorbed Fe^{3+} in chlorite and illite.

In a later study, Koppelman, et al. (63) studied Cr^{3+} adsorption on chlorite, illite, and kalonite over a pH range of 1–10. They found that from pH 2 to 4 the $Cr(2p_{3/2})$ binding energy (~577.3 eV) was in agreement with their earlier study (Table 3). At higher pHs, the binding energy (~576.8 eV) corresponded to $Cr(OH)_3$. They concluded that below pH 5, the Cr^{3+} was adsorbed as the aquated ion $Cr(H_2O)_6^{3+}$, while at higher pHs the Cr^{3+} was precipitated as $Cr(OH)_3$ in agreement with solubility predictions.

Koppelman and Dillard have also studied the adsorption of Ni^{2+} (64), Cu^{2+} (64), and Co^{2+} (65) on chlorite. By comparing binding energies for the adsorbed $Ni(2p_{3/2})$, $Cu(2p_{3/2})$, and $Co(2p_{3/2})$ peaks with standard metal oxides, hydroxides and minerals, they concluded that Ni^{2+} was probably adsorbed as the aquated ion $Ni(H_2O)_6^{2+}$ at pH 6, that Cu^{2+} may be adsorbed as the hydrolyzed $Cu(OH)^+$ at pH 5, and that Co^{2+} was probably adsorbed as the aquated ion $Co(H_2O)_6^{2+}$ at pH 3 and 7.

The adsorption of Co^{2+} on hydrous manganese dioxide, a disordered birnessite, has been studied at several pHs below 7.1 (66). The XPS spectrum suggested that Co^{2+} was oxidized to Co^{3+} on adsorption. Evidence for this (Figure 7) was found in the lower binding energies for the $Co(2p^3)$ and $(2p^1)$ peaks in Co^{3+} compounds, the smaller binding energy difference between the $Co(2p^3)$ and $(2p^1)$ peaks in Co^{3+} compounds, and the absence of both the prominent shake-up satellite 5 or 6 eV above the $Co(2p_{3/2})$ peak and the broad peak widths typically found in paramagnetic Co^{2+} compounds.

XPS has been used to study the kinetics of adsorption for Ni^{2+}, La^{3+} and Ba^{2+} at pH 8.3 on thin films of amorphous hydrous manganese dioxide (67). Using XPS-derived calibration plots, a selectivity sequence of $Ba^{2+} \gg Ni^{2+} \gtrsim La^{3+}$ was determined in agreement with the average metal content of deep-ocean sediments (68). XPS-determined adsorption plateaus corresponded to monolayer coverage or greater. Based on the binding energies of the $Ni(2p_{3/2})$ peak and its shake-up satellite, the authors speculated that Ni^{2+} was present as Ni_2O_3 or $Ni(OH)_2$ even though Ni concentrations never exceeded solubility product constraints.

XPS-derived calibration curves have also been used to study Ba^{2+} adsorption on single crystals of calcite (69). Ba^{2+} adsorption curves corresponded to a maximum coverage of approximately one monolayer. No information on the nature of the adsorbed Ba^{2+} was reported.

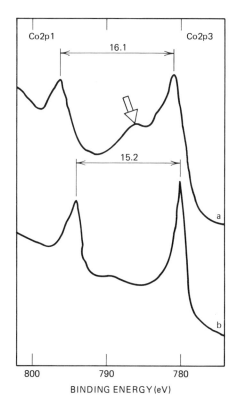

Figure 7. Cobalt $2p$ XPS data for (a) Co(OH); (b) Co^{2+} adsorbed on MnO_2. Arrow points to shake-up satellite in the Co(OH)$_2$ spectrum. Compiled from data cited in Murray and Dillard (66).

Brown et al. (70) evaluated the utility of the iron sulfide minerals FeS (pyrrhotite) and FeS_2 (pyrite) as adsorbing surfaces for Hg^{2+} and Hg^0 in mercury pollution control. Their results showed that Hg^{2+} adsorption on FeS at pH 4 increased with an increase in Cl^- concentration, although little Cl^- was detected by XPS. Adsorption rates were found to be lower at higher pHs and on FeS_2.

The adsorption of Co^{2+} on Al_2O_3 and ZrO_2 has been studied (71). The comparison of $Co(2p_{3/2})$ binding energies and shake-up satellite structure showed that at 30°C, Co^{2+} was adsorbed as Co(OH)$_2$ on both Al_2O_3 and ZrO_2, while at 200°C, the Co^{2+} adsorbed on Al_2O_3 corresponded to $CoAl_2O_4$, suggesting modification of the alumina surface. Unlike a previous study of Co^{2+} adsorption on MnO_2 (66), the Co^{2+} adsorbed on Al_2O_3 and ZrO_2 was not oxidized to Co^{3+}.

Alvarez et al. (72) have studied the adsorption of Si(OH)$_4$, CaSiO$_3$, and Ca(H$_2$PO$_4$)$_2$ on Al(OH)$_3$. When the Al(OH)$_3$ was treated with Si(OH)$_4$

solutions, no Si was detected by XPS. Treatment with $CaSiO_3$ and $Ca(H_2PO_4)_2$ solutions, however, gave Ca and Si, and Ca and P XPS peaks, respectively. The authors concluded that silicate adsorption on $Al(OH)_3$ could not occur in the absence of Ca. No XPS evidence was found for the surface modification of $Al(OH)_3$ following adsorption.

The adsorption of metal–ligand complexes on clay minerals has also been investigated. XPS results for $Cr(NH_3)_6^{3+}$ and $Cr(en)_3^{3+}$ adsorbed on chlorite at pH 3 are shown in Table 4 (73). The binding energy values for the $Cr(2p_{3/2})$ peaks were in agreement with results for the adsorption of aquated Cr^{3+} (see Table 3). The amine N/Cr ratio suggests that the adsorbed complexes were at least partially hydrolyzed to aquo complexes. It was not possible, however, to determine if the adsorbed species were $Cr(H_2O)_6^{3+}$ or partially hydrolyzed Cr^{3+}–ligand complexes. Binding energies and satellite structures showed that $Co(NH_3)_6^{3+}$ was reduced to Co^{2+} after adsorption onto chlorite at pH 3 (65).

3.3. Auger and Ion Microprobe Analysis

The most extensive environmental application of the Auger and ion microprobe techniques has been to the characterization of pollutant particles produced in the combustion of fossil fuels (74–88). Of particular interest has been the characterization of particulate by-products generated by

TABLE 4.   XPS Results for Cr^{3+}-Ligand Adsorption on Clay Minerals[a]

$Cr(NH_3)_6^{3+}$ on Clay	$Cr(2p_{3/2})$ (eV)	Amine N/Cr
Chlorite	577.2	3.2
Illite	577.2	2.8
Kalonite	577.3	3.8

$Cr(en)_3^{3+}$ on Clay	$Cr(2p_{3/2})$ (eV)	Amine N/Cr
Chlorite	577.4	2.1
Illite	577.4	4.2
Kalonite	577.4	4.1

[a] Data from Reference 73.

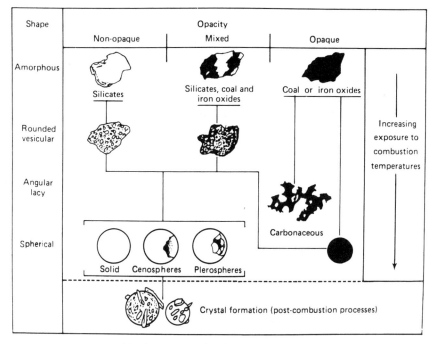

Figure 8. Fly ash classification scheme showing relationship of morphology to combustion temperature (83).

various coal utilization technologies that may be increasingly utilized to help offset anticipated energy deficits in the United States (83). The combustion of coal produces solid wastes composed of noncombustible minerals, as well as organic or carbon containing particles that result from varying degrees of combustion efficiency. Of major interest is fly ash, the portion of the solid waste particulates that is of sufficiently small diameter to be easily suspended in the flue gas and to penetrate through conventional emission control systems into the atmospheric environment. Total ash production in the United States is projected to reach 125 million tons by 1990 (89). An excellent review of all aspects of coal fly ash including physical, chemical, and environmental factors has recently been published (83).

Coal fly ash is a complex mixture containing particles of varying size, morphology, and chemical composition (Figure 8) (83, 90). Specific research interest in the surface chemistry and physics of fly ash particles was first stimulated by studies of trace element bulk concentrations (μg/

g basis) as a function of particle size. Initial studies by Davison et al. (91) indicated that a large number of potentially toxic trace elements (Pb, Tl, Sb, Cd, As, Se, Cr, Ni, S) increased in bulk concentration with decreasing particle size. The elemental partitioning mechanism proposed involved volatilization of certain elements or their compounds in the high-temperature combustion zone and their subsequent condensation or adsorption on entrained particles in the cooler regions of the exhaust train. The larger surface area to volume ratios of smaller particles would explain the size dependence of bulk concentrations. A variety of subsequent studies (92–97) have confirmed the presence of enhanced concentrations of many elements (As, B, Ba, Br, Cd, Cl, Co, Cr, Cu, Ga, Hg, I, Mg, Mn, Mo, Ni, Pb, Po, Rb, S, Sb, Sc, Se, Sr, Tl, W, V, Zn) in the smaller fly ash size fractions. Similarly, bulk concentrations of many elements in fly ashes discharged into the atmosphere are enhanced relative to the progressively decreasing bulk concentrations observed in fly ash collected in the stack by electrostatic precipitators, in bottom ash or slag, or in the original coal sample, respectively (92–95). Mathematical models for enrichment via the proposed vaporization–condensation mechanism also have been developed (92, 97).

The environmental significance of the preceding findings is directly related to particle surface chemistry. The condensation of volatile and potentially toxic species on particle surfaces predicts enhanced bulk concentrations in the smaller particles, which most readily escape control devices, have low gravitational settling velocities resulting in long atmospheric residence times, and preferentially deposit in pulmonary lung regions (74, 76–79, 83, 91, 98–100). The actual concentrations at the point of contact between an inhaled particle and living tissue also would be substantially higher than suggested by bulk chemical measurements.

Direct evidence for trace element surface predominance in coal fly ash has been obtained via depth profiling measurements of single particles using ion sputtering in concert with Auger and ion microprobe techniques (74–87). Positive correlations usually are observed between trace elements exhibiting an inverse dependence of bulk concentration on particle size and those with enhanced surface region concentrations (e.g., Pb, Tl, Cr, Mn, V) (74, 76–78, 80, 82, 83). The extent of surface predominance and thickness of surface predominant layers may vary considerably between various types of coal combustion devices and the nature and location of particle collection schemes used to obtain analytical samples.

Typical surface layer thicknesses, however, are on the order of 100–1000 Å. More refractory metals incorporated as oxides in the glassy fly ash matrix (e.g., Al, Si, and Ti) exhibit little particle size dependence or surface concentration enrichments (74, 76–78, 80, 82–84).

A substantial body of evidence derived from surface analysis measurements (SIMS, Auger electron spectroscopy, XPS) suggests that the inorganic components of the surface layers are largely metal and ammonium sulfates (74, 76–78, 80, 82–84, 86, 101, 102). The presence of acid sulfate layers is of importance with respect to the role of fly ash particle surfaces in the heterogenous catalytic oxidation of sorbed sulfur oxides (103). Particle surfaces rich in sulfuric acid also may promote the condensation of certain trace elements (104) or result in the post-combustion crystallization of surface sulfate compounds (90, 105). The presence of major amounts of alkali and alkaline earth metals (Na, K, Ca) and sulfate on fly ash surfaces is very similar to the composition of condensed corrosion deposits found on power plant boiler surfaces (78). Dis-· solution of these various acid sulfate species in raindrops or in aquatic systems following dry deposition contribute to environmental problems associated with acid rain or increased acid loadings of natural waters.

Indeed, solvent leaching studies have demonstrated that surface region components are preferentially soluble in comparison to elements contained primarily in the aluminosilicate matrix of fly ash (78, 82, 83). Ion microprobe depth profiles obtained after solvent leaching also show that surface layers enriched in trace elements have been largely removed (78, 80). Hence surface species are potentially available to aquatic ecosystems or to body fluids or cells following particle inhalation. With regard to the latter, ion microprobe experiments have been initiated to directly examine the role of particle surface composition in cellular toxicology (106). Cell cultures (alveolar macrophages) exposed to Pb_3O_4 particles or lead-oxide-coated fly ash have been examined using correlative ion and electron microanalysis (106). Lead toxicity was manifest in that dissolution, and migration of lead from particle surfaces resulted in the formation of Pb-, Ca-, and P-containing precipitates (106, 107).

Mutagenic, volatile organic compounds including polynuclear aromatic hydrocarbons and their derivatives also may be enriched on fly ash particle surfaces as the result of sorption processes similar to those described earlier. The PAS technique is more suited to the characterization of sorbed organic species, as is described in other sections of this chapter. Further details of Auger and ion microprobe studies of coal fly ash are described in Section 4.

Results comparable to those of the coal fly ash surface studies have been obtained for automobile exhaust particles produced in the combustion of leaded gasoline (76, 80, 108). Elements associated with volatile lead compounds (Pb, Br, Cl) are surface predominant on large (>5 μm cross section), iron-rich refractory particles produced primarily by ablation or corrosion of the exhaust system. XRD studies indicate that the major crystalline form of Pb is $PbSO_4$ (108, 109). Auger microprobe depth profiles (80) also indicate very large surface region sulfur concentrations, although XPS studies have not been conducted to verify surface sulfate speciation. The transformation of Pb species containing halogens to sulfates is consistent with Br and Cl loss mechanisms observed for aging automobile exhaust particles and with the identification of $PbSO_4$ as the major crystalline form of Pb in urban dust samples contaminated with automobile exhaust particles (108, 110). The volatilization–condensation mechanism resulting in toxic metal surface predominance and the ultimate formation of crystalline sulfate surface layers appears virtually identical to that observed for coal fly ash.

Correlative XPS, Auger electron spectroscopy, ion microprobe, and time-resolved solvent leaching studies also have been conducted on particles derived from open hearth blast furnaces (111). Although extensive trace element surface predominance is not observed, surface layers enriched in acid sulfates are present, analogous to the results for coal fly ash and automobile exhaust particles. Further elaboration of these results can be found in Section 4.

IMP mass spectrometry also has been used to characterize individual oil soot particles (2–8 μm cross-sectional diameters) (81). Although no elemental depth profiling was attempted, ion micrographs showing interparticle and intraparticle distributions of selected elements were obtained. Quantitative analyses were reported using elemental sensitivity factors derived from oxide, glass, and mineral standards. Oil soot could be readily identified in the presence of other air pollution particles on the basis of large V concentrations (typically 70,000–150,000 μg/g). This application points to the possibility of using IMP techniques to identify specific sources of pollutant particles in environmental samples through characteristic individual particle compositions. Such specificity is sacrificed in conventional bulk determinations on a collection of particles. In comparison to Auger or electron microprobe based source attribution studies, the ion microprobe technique offers specific advantages of generally higher elemental sensitivity and isotopic detection capability. For example, various particulate sources of Pb could perhaps be distinguished by SIMS on the basis of unique Pb isotope ratios (108). Similarly, radioactive species in respirable particles (e.g., ^{214}Po, ^{210}Po, ^{210}Pb) could be characterized in studies related to bronchial cancers (112, 113). Positive and negative secondary ion spectra obtained from individual particles in urban air reveal the potential of SIMS for the characterization of chemically complex, multielemental samples (47).

4. COMPARISON OF SURFACE TECHNIQUES IN ENVIRONMENTAL ANALYTICAL CHEMISTRY

4.1. Elemental Sensitivity

Much of the recent interest in environmental chemistry and toxicology is a direct result of the evolution of analytical tools with sufficient sensitivity to permit monitoring of low doses of toxic materials. The accelerated exposure to toxic trace elements is the direct consequence of their increased mobilization in the biosphere resulting from raw materials acquisition and the energy consumption required to sustain "advanced" technologies. The sudden nature of the mobilization is of ecological significance in that organisms do not have the benefit of evolutionary mechanisms to provide effective homeostatic defenses against these xenobiotic materials (80).

Of the common surface techniques applied to elemental analysis, SIMS typically provides the best elemental detection limits (Figure 9). However, the enhanced sensitivity is achieved at the expense of major variations in sensitivity as a function of atomic number. The periodic nature of SIMS sensitivity shown in Figure 9 is typical of that observed using rare gas or oxygen ion bombardment and positive secondary ion detection (114). Elements with high electron affinity are more sensitively detected using an electropositive primary ion (Cs^+) and negative secondary ion detection (115). Therefore, instrumentation providing variable primary ion composition and both positive and negative ion mass spectrometry permits SIMS detection limits for most elements to be realized in the low parts-

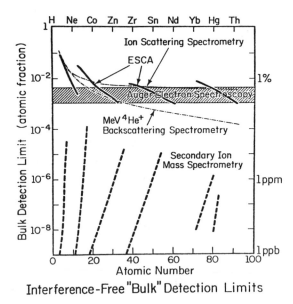

Interference-Free "Bulk" Detection Limits

Figure 9. Comparison of detection limits of various surface techniques (SIMS data assumes positive secondary ion detection). Reprinted with permission from C. A. Evans, Jr., *Anal. Chem.* **47**(9), 818A (1975). Copyright 1975 American Chemical Society.

per-million range or below. The enhanced elemental detection limits of SIMS in comparison to XPS and Auger electron spectroscopy has been illustrated for environmental samples such as coal fly ash particles (Table 5) (78, 82). Only SIMS is capable of routine detection of elements with surface region concentrations well below 1 at. %.

4.2. Spatial Resolution

4.2.1. Lateral Resolution

The heterogeneous nature of environmental particulate samples requires the analytical specificity of single particle analysis, particularly in studies related to the tracing or source attribution of toxic species (108). In the atmospheric environment, single particles with diameters on the order of a few microns or less are of particular interest in that they are least effectively collected by existing control technologies, have long atmospheric residence times, may be enriched in toxic species in comparison

TABLE 5. Surface Predominance of Elements in Coal Fly Ash Particles[a]

Element	Bulk Concentration[b] ($\mu g/g$)	$\dfrac{\text{Signal Intensity at Surface}}{\text{Signal Intensity at 500-Å Depth}}$	Tech-nique[c]
Matrix Elements			
Al	>15,500	1.1	a, b, c
Fe	92,400	0.7	a, b, c
Si	110,000	1.1	a, b, c
Minor Elements			
C	—	3.5	a, b, c
Ca	>28,600	1.6	a, b, c
K	38,800	7.6	a, b, c
Mg	12,300	0.9	a
Na	>19,700	15.2	a, b, c
S	7,100	7.7	a, b, c
Ti	4,700	0.9	a
Selected Trace Elements			
Be	32	6.0	a
Cr	380	3.3	a
Li	200	3.8	a
Mn	310	6.4	a
P	600	3.8	a
Pb	620	11.0	a
Tl	28	10.0	a
V	380	2.0	a
Zn	1,250	7.2	a

[a] Data from References 82 and 109.
[b] Determined by spark source mass spectrometry.
[c] The surface techniques capable of detection of surface elements: a = ion microprobe mass spectrometry; b = Auger electron spectrometry; c = x-ray photoelectron spectroscopy.

161

to larger particles, and preferentially deposit in pulmonary regions of the lung, where toxicologic impact may be most severe (see Section 3 of this chapter).

Recent advances in electron and ion optics permit the extention of Auger and ion microprobe techniques to single particles having submicron dimensions. Several Auger microprobes using field emission sources that are now in use provide primary electron beam diameters on the order of 500 Å (e.g., JEOL, Perkin-Elmer, Physical Electronics Industries). The high current densities usually associated with these microbeam sources may, in fact, help to reduce charging effects typically observed for non-conducting environmental particles (85). However, the high localized doses of primary electrons also may give rise to a variety of potential artifacts, including electron stimulated surface desorption, surface oxidation or reduction, and field enhanced or thermal diffusion of mobile ions (e.g., alkali metals in fly ash) (85, 86, 109, 116).

The second generation ion microscope (Cameca IMS-3f) permits lateral resolutions approaching 0.3 μm to be achieved for ideal samples (117). Single toxic particles of micron dimensions (including aluminosilicates and Pb_3O_4) have been observed in biological samples including lung tissue (118) and cultures of alveolar macrophages grown as monolayers on metal surfaces (106). Ion microprobe measurements of single particles with edge resolutions approaching 1 μm also have been accomplished (47). A limitation of the IMP configuration is that the very high microprobe current density usually causes increases in both sputtering rate (resulting in possible reductions in particle sputter lifetime or the depth resolution of multielemental depth profiles) and possible charging effects (resulting in aberrations in secondary ion extraction fields and reduced lateral resolution). An advantage of the ion microscope configuration is that a broad primary beam (~50–100 μm) can be used with consequent reductions in current density, while lateral resolutions on the order of 1 μm can be maintained (117).

Both the PAS and XPS techniques are considerably more limited with respect to lateral resolution. Although not widely available, a scanning XPS system proposed by Cazaux (119) and utilized by Hovland (120) could be utilized for single particle analysis with a spatial resolution approaching 10 μm. The sensitivity requirements of PAS generally demand that rather large sample surface areas be irradiated even when intense laser sources are available. Since typical photoacoustic signals originate from surface temperature variations of less than 0.001°C (121), the effective size of the photoacoustic piston is quite small, thereby requiring the presence of many photoacoustically active chromophores within rather large analytical volumes.

4.2.2. Depth Resolution

Since the unique chemical nature of particle surfaces cannot be fully appreciated without compositional information about subsurface layers, it is imperative that the surface techniques have the additional capability of depth profiling analysis.

The sampling depth in Auger electron spectroscopy or XPS is dependent on the inelastic mean free path (IMFP) of the Auger or x-ray photoelectrons which are, in turn, dependent on kinetic energy (E) (Figure 10). The IMFP values can be predicted from a least-squares fit of the data in Figure 10:

$$\lambda_i = \frac{A_i}{E^2} + B_i E^{1/2} \tag{3}$$

where λ_i is the IMFP in nanometers, monolayers, or mg/m^2. A_i and B_i are constants that are dependent on both the matrix (elements, inorganics, organics or adsorbed gases) and the units of λ_i (122). The IMFP for a 1-

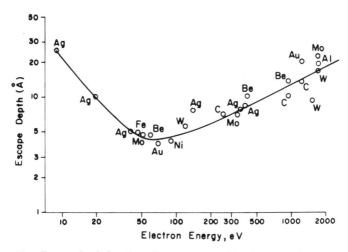

Figure 10. Escape depth for photoelectrons versus kinetic energy for metals (30).

keV electron is on the order of 5–20 Å in metals, 15–40 Å in metal oxides, and 40–100 Å in polymers (30).

By decreasing the takeoff angle for Auger or x-ray photoelectrons entering the energy analyzer (Figure 11), the effective sampling depth can be decreased. The result is to increase the surface sensitivity of XPS or Auger electron spectroscopy (30), thus providing a nondestructive technique for the depth-profiling of near-surface layers. Grazing angle XPS measurements have been used in the study of Co^{2+} adsorption on chlorite and illite (123). The intensity enhancement ratio:

$$\frac{[Co(2p_{3/2})/Si(2s)]_{11°}}{[(Co(2p_{3/2})/Si(2s)]_{90°}} \tag{4}$$

was greater than 1.0, suggesting that the Co^{2+} was adsorbed on the clay surfaces. In general, the surface roughness, porosity, and nonplanar geometry of single particles limits the utility of angular resolved XPS or Auger electron spectroscopy measurements.

A more general approach to depth profiling involves ion sputtering to remove surface layers of material. This is required in SIMS as an integral part of analytical signal generation, but also may be used in conjunction with XPS/Auger electron spectroscopy to provide analyses of subsurface layers revealed by ion sputtering. Differentially pumped ion guns also permit both XPS and Auger electron spectroscopy data acquisition to proceed simultaneously with ion sputtering. Quantitative depth profiling

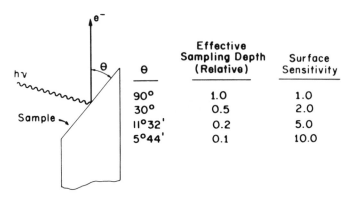

θ	Effective Sampling Depth (Relative)	Surface Sensitivity
90°	1.0	1.0
30°	0.5	2.0
11°32'	0.2	5.0
5°44'	0.1	10.0

Figure 11. Effective sampling depth and surface sensitivity for various photoelectron takeoff angles (30).

may be limited by sputtering-induced alterations in subsurface composition including cascade mixing and preferential sputtering effects (124, 125) and by particle geometry and porosity resulting in nonuniform sputtering and analytical sampling depths (126). Auger microprobe depth profiles of small single particles mounted on metal substrates (sputtering using broad ion beams) also may be limited by sputter deposition of the metal substrate material on the particle surfaces (87). Thin film metal oxide standards may be used for calibration of nominal sputtering rates, which are on the order of 1 Å/sec using rare gas ion sources typically available with Auger electron spectroscopy/XPS instrumentation (78, 87). Reactive gas ion sources of higher current density available with most ion microprobes permit much higher sputtering rates.

Auger microprobe depth profiles of isolated cenospheres (88) produced in fluidized bed coal combustion provide evidence for an alternating layered structure as illustrated in Figure 12. IMP depth profiles of single ash particles derived from conventional coal combustion (Table 5) provide detailed evidence of trace element surface predominance resulting primarily from volatilization–condensation processes occurring in the combustion zone and stack, respectively (see Section 3 of this chapter).

Although not capable of the depth resolution provided by XPS, Auger electron spectroscopy, or SIMS techniques, PAS may provide information on submonolayer coverages of adsorbed materials (in favorable cases). PAS instruments also provide the means to experimentally vary analytical sampling depths (analogous to variable angle XPS or Auger electron spectroscopy measurements). According to a model proposed by Rosencwaig and Gersho [RG theory (9, 10)], heat flow within the sample is both diffusion and modulation frequency dependent. Since heat generated deep within the sample will not reach the surface before the next pulse from a modulated source, the radiationless relaxation from these interior chromophores appear as steady state heat flow rather than as a modulated thermal wave. By increasing the modulation frequency, the analytical sampling depth is reduced (thermal diffusion length, μ_S). If $\mu_S > \mu_\beta$ (instantaneous depth of optical penetration), the photoacoustic signal is no longer indicative of analyte concentration. This condition is referred to as photoacoustic saturation. However, if the modulation frequency (ω) is increased, μ_S is reduced, thereby decreasing the depth of analysis. In some cases, ω may be increased to avoid saturation. However, thermal diffusion length, μ_S, varies as $\omega^{-1/2}$, and the photoacoustic

a) Aluminum & Silicon

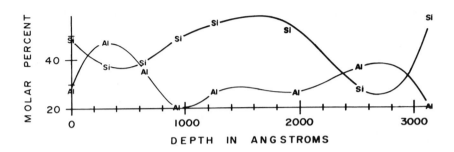

b) Potassium & Sulfur

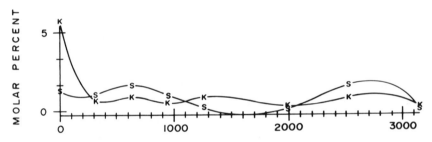

Figure 12. Elemental depth profiles of a single particle obtained from a fluidized bed coal combustor and characterized using Ar^+ sputtering and scanning Auger microanalysis (88).

signal magnitude varies as $\omega^{-3/2}$ (for materials not saturated) or ω^{-1} (for saturated materials). Therefore, increasing ω greatly reduces sensitivity and limits the usefulness of depth-profiling analyses based on such techniques. Piezoelectric detectors may be used, however, to increase sensitivity for samples that make good thermal contact with the detector.

Diffusion controlled heat flow within the thermal diffusion length causes a phase (time) lag of the photoacoustic response with respect to the excitation. This phase lag is indicative of the depth of the photoacoustically active chromophore within the sample. Therefore, phase sensitive lock-in amplification of the photoacoustic response also may be used to depth profile the sample (127, 128). However, these depth profiles have typical depth resolutions on the order of tens of micrometers (128) rather than nanometers as in Auger electron spectroscopy and XPS. Because of the complicated phase relationships applicable to particulate samples, this method currently is not very quantitative for such materials.

A promising application of PAS as a surface probe in environmental chemistry is the characterization of photoacoustically active analytes sorbed on particulate substrates, for example, combustion-generated aerosols having sorbed organic materials including polynuclear aromatic hydrocarbons and their derivatives (129–132). Figure 13 (51) shows the UV photoacoustic spectrum of (a) fluorene vapor adsorbed on silica gel, (b) untreated silica gel, and (c) the result of a digital subtraction of spectrum b from spectrum a. The surface coverage was ≤4 mg of fluorene per gram of silica. This corresponds to 0.2 of a monolayer surface coverage, thus illustrating the surface sensitivity potentially available via PAS (51).

4.3. Chemical Speciation Information

Although a vast array of trace elemental analysis techniques are available, there are relatively few analytical tools capable of providing *in situ* information on the chemical speciation of individual elements present in particulate sample matrices. Speciation questions have a direct bearing on the potential interactions of individual pollutants with other environmental constituents, on phase transitions to or from the particulate state in various environmental compartments, and on biological availability and toxicity (76, 80, 133).

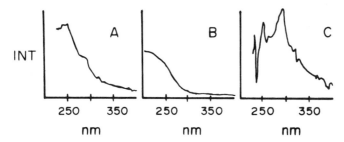

Figure 13. PAS studies of fluorene adsorbed from the vapor phase on silica: (*a*) sample/reference; (*b*) untreated silica (blank); (*c*) sample-blank/reference. Reference is a carbon black spectrum used to provide source compensation. Reprinted with permission from Cabaniss and Linton (51).

In XPS, the spectrometer's energy scale must be calibrated before the measured kinetic energies can be used for chemical speciation information. Nonconducting samples usually develop a positive charge during x-ray irradiation that retards the ejected photoelectrons and leads to higher apparent electron binding energies. If accurate binding energy information is to be obtained, a suitable means for charge referencing must be employed. Methods include gold decoration, use of the adventitious carbon (1s) line and vapor deposition of a hydrocarbon (134).

In some cases charge referencing to the C(1s) peak is difficult due to either low intensity or a broad peak width. When gold decoration or vapor deposition of a hydrocarbon are not practical, charge corrections can be referenced to the binding energy of a photopeak line from the sample matrix. Zn($2p_{3/2}$) XPS data are shown in Figure 14 for zinc adsorption on hydrous ferric oxide (HFO) as a function of pH. Charge corrections were made by referencing to the Fe($2p_{3/2}$) binding energy which is independent of solution pH (Table 6). The average Fe($2p_{3/2}$) binding energy of 711.5 eV is in good agreement with a previously determined value of 711.6 eV (58). Changes in the Zn($2p_{3/2}$) peak position and line shape represent a significant change in zinc speciation as a function of pH (135).

In some samples the XPS lines for different elemental chemical states are partially or totally unresolved. The interpretation of incompletely resolved XPS data can be facilitated with the application of curve-fitting

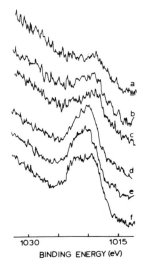

1030   1015
BINDING ENERGY (eV)

Figure 14. Zinc ($2p^3$) XPS data for Zn adsorbed on hydrous ferric oxide (HFO) as a function of solution pH: (*a*) pH = 6.00; (*b*) pH = 6.45; (*c*) pH = 6.70; (*d*) pH = 6.80; (*e*) pH = 7.20; (*f*) pH = 8.40. Initial solution concentrations were [Zn] = 1 × 10^{-4} M, [HFO] = 1 × 10^{-3} M (D. T. Harvey and R. W. Linton, unpublished data, 1981).

TABLE 6. Hydrous Ferric Oxide
Fe($2p_{3/2}$) Binding Energies[a]

pH	Fe($2p_{3/2}$) (eV)
6.20	711.4
6.25	711.3
6.60	711.8
7.30	711.3
7.40	711.7
8.30	711.5
Average	711.5 ± 0.2

[a] D. T. Harvey and R. W. Linton, unpublished data, 1981.

routines. Individual peaks may be defined (Figure 15) by their height (H), width at half-height (W), binding energy of peak maximum (P), Gaussian–Lorentzian fraction and a sigmoidal curve to account for energy loss features (136). Harvey and Linton (58) used curve fitting of the oxygen ($1s$) peak to determine the oxide, hydroxide, and physically adsorbed water content of several iron oxide samples (Figure 16, Table 7). Binding energy values for the oxide and hydroxide oxygen lines were found to be in good agreement with previously determined values (137, 139).

Since the Auger process involves valence electrons, the chemical shifts for x-ray excited Auger lines may be larger than the XPS lines involving the same core-level vacancy. Plots of the Auger kinetic energy versus the

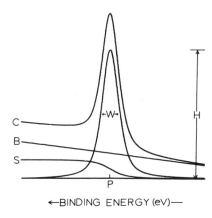

Figure 15. Calculated spectrum (C) used in XPS curve fitting. P = XPS peak position; W = XPS peak width; H = XPS peak height; B = baseline; S = sigmoidal curve.

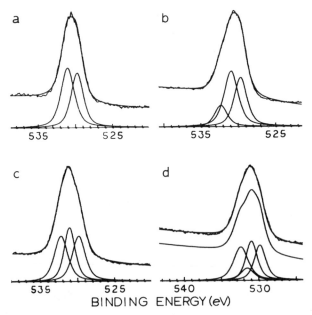

Figure 16. Oxygen ($1s$) curve-fitting results for: (a) goethite; (b) aged hydrous ferric oxide (HFO); (c) limonite; (d) fresh HFO (the extra curve accounts for a minor O contribution from the Al foil substrate used to mount the sample). Reprinted with permission from D. T. Harvey and R. W. Linton, *Anal. Chem.* **53**, 1689 (1981). Copyright 1981 American Chemical Society.

XPS binding energy have proven useful in the identification of surface chemical states. Figure 17 shows the results of a study of several iron oxide samples (58). The kinetic energy of the oxygen KVV peak decreases and the binding energy of the oxygen ($1s$) peak increases with an increase in the amount of physically adsorbed water present (Table 7). The binding energy for the fresh HFO sample appears out of order due to a broader peak width for physically adsorbed water. The diagonal lines are the modified Auger parameters, which are independent of surface charging (134). Wagner et al. have published extensive Auger parameter data for 24 different elements (134) and oxygen in oxides, hydroxides, anions, and carbon–oxygen compounds (140).

Because of the complex nature of the multielectron transition leading to ultimate Auger electron production, changes in electron-excited Auger line shapes and energy usually are less readily interpreted with respect to surface chemical state changes than XPS chemical shifts. However,

TABLE 7. Results of Spectral Curve Fitting of Iron Oxide Oxygen (1s) XPS Peaks[a]

	Binding Energies (eV)			Spectral Intensities		
				Hydroxide	Physically Adsorbed H_2O	
Sample	Oxide	Hydroxide	Physically Adsorbed H_2O	Oxide	Oxide	n value for $Fe_2O_3 \cdot nH_2O$[b]
Goethite	529.6 ± 0.1	530.9 ± 0.1	—	1.06 ± 0.13	—	1.04
Aged HFO[c]	529.6 ± 0.1	530.9 ± 0.1	532.3 ± 0.1	1.10 ± 0.20	0.37 ± 0.05	1.43
Limonite	529.8 ± 0.1	531.0 ± 0.1	532.2 ± 0.1	1.00 ± 0.10	0.96 ± 0.20	1.96
Fresh HFO[c]	529.8 ± 0.1	530.9 ± 0.1	532.3 ± 0.1	1.17 ± 0.11	1.13 ± 0.12	2.24

[a] Data from Reference 58.
[b] Chemically + physically adsorbed water.
[c] Hydrous ferric oxide (HFO).

171

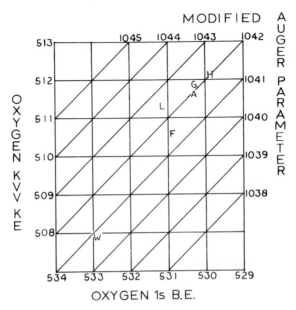

Figure 17. Oxygen chemical state plot: H = hematite; G = goethite; A = aged HFO; L = limonite; F = fresh HFO; W = water. Reprinted with permission from D. T. Harvey and R. W. Linton, *Anal. Chem.* **53**, 1689 (1981). Copyright 1981 American Chemical Society.

empirical correlations have been obtained, for example for various particulate metal–sulfur compounds, showing rather substantial variations in metal and sulfur Auger line shapes (141). Auger transitions involving valence shells offer particular promise for the evolution of improved Auger line-shape analysis for chemical state information (142).

The inherently destructive nature of high primary ion beam current densities typical in ion microprobe versions of SIMS limits its utility as a speciation tool. However, the use of large primary beams with low current density allows highly surface specific measurements in which sputter rates are typically 10^{-3} atomic layers/sec. The so-called static SIMS experiment is useful for elucidation of surface structure and speciation via analysis of mass fragmentation patterns (143). For example, the technique may be a viable approach to characterize the chemistry of single organic compounds (e.g., aromatic hydrocarbons) sorbed on particle surfaces as models for the study of heterogeneous atmospheric reactions. Organic "fingerprint" mass clusters have been observed in preliminary static SIMS experiments involving diesel exhaust particles suggesting the presence of residual organics not removed by extracting solvents (144).

The coupling of solvent extractions and depth profiling experiments involving surface analysis techniques provide important speciation information with regard to the "availability" of elements in environmental particulates. The availability of an element is related to both the accessibility of the extracting solvent to the region where it is present and to the solubility of the chemical species in which the element exists. Bulk leaching studies of coal fly ash suggest a strong positive correlation of high surface accessibility determined by SIMS and high extractability (Table 8) (78, 82, 83).

Similar ion microprobe studies of dusts derived from open hearth blast furnaces reveal minimal surface predominance of most potentially toxic metals (e.g., Mn, Cr, Zn, Pb) (111). However, time-resolved leaching studies (showing incremental amounts of elements leached versus time) show initial "bursts" of leached metals (Figure 18) and sulfate. Depth profiles of sulfate by XPS reveal extensive surface predominance (111). Despite the lack of metal surface enrichments, these results offer evidence for the surface transformation of sorbed sulfur species to soluble metal sulfates. The coupling of both surface analysis and time-resolved leaching experiments potentially permits, therefore, the evaluation of the surface solubility, accessibility, and speciation of potentially toxic materials associated with environmental particulates.

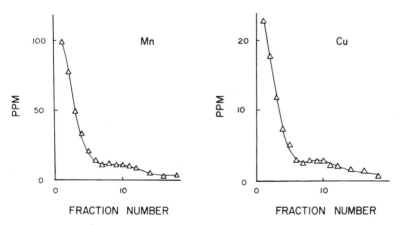

Figure 18. Time-resolved water leaching of open hearth blast furnace dust. Solution concentrations are shown for successive 5-mL fractions (111).

TABLE 8. Correlation of Surface Analysis and Solvent Extraction Studies of Coal Fly Ash[a]

Elements	Bulk Concentration (μg/g)[b]	Average Concentration in 1000-Å Surface Layer[a] (μg/g)[c]	Percent Extracted in H_2O
Matrix			
Fe	92,400	84,000	0.7
Si	110,000	21,000	0.1
Minor			
Ca	12,500	21,000	1.3
K	38,800	103,000	6.0
Mg	12,300	9,800	0.6
Na	5,200	35,600	5.2
S	7,100	252,000	36.0
Ti	4,700	9,600	1.5
Trace			
Cd	24	700	19.0
Cr	400	1,400	5.7
Li	200	4,400	17.0
Mn	310	2,900	35.0
P	600	4,100	11.0
Pb	620	2,700	3.4
Tl	28	920	25.0
V	380	760	3.9
Zn	1,250	14,600	22.0

[a] Data from References 82 and 109.
[b] Determined by spark source mass spectrometry.
[c] Determined by IMP mass spectrometry.

The utility of solvent extractions extends to the analysis of polynuclear aromatic hydrocarbons sorbed on environmental particulates (145). Chromatographic separation and quantitation permits the characterization of complex mixtures, but suffers from variations in solvent extraction efficiencies for various extract components. Removal of the sorbed material from the substrate also destroys any sorbent–sorbate interaction information, which is a key to understanding the environmental chemistry of organic vapor–particle interactions.

Cabaniss and Linton (51) have attempted to model particulate material–organic interactions using PAS to characterize substrates with vapor-deposited polynuclear aromatic hydrocarbons. Figure 19 shows the UV–PAS spectra of fluorene and fluorenone on silica gel. These photoacoustic spectra resemble the dilute solution UV–VIS absorption spectra (Figure 19), but the similarity of these spectra also illustrates the limited chemical speciation capabilities of UV–PAS. The oxidation of fluorene to fluorenone on fly ash particles has been documented using solvent extraction techniques (131). Improved speciation information available using FT–IR–PAS is illustrated in Figure 20, which compares the conventional IR spectra of PAH's to FT–IR–PAS spectra. Sensitivity considerations generally result in long periods of signal averaging and purging of the photoacoustic cell with He gas.

4.4. Quantitation of Surface Concentrations

Empirical approaches to quantitation dominate the current application of surface techniques to environmental particulate characterization. Relative elemental sensitivity factor approaches in XPS, Auger electron spectroscopy, and SIMS provide concentrations with uncertainties generally

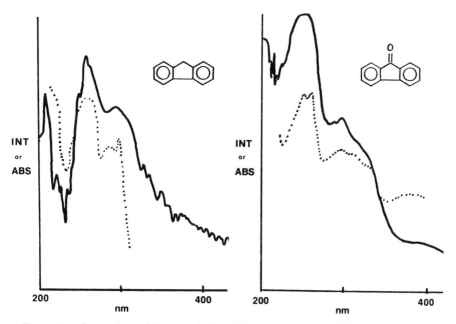

Figure 19. Comparison of fluorene (left) and fluorenone (right): UV-PAS (solid line) of solution adsorbed PAH species on silica; UV absorption spectra (dotted line) of dilute PAH species in solution. Reprinted with permission from Cabaniss and Linton (51).

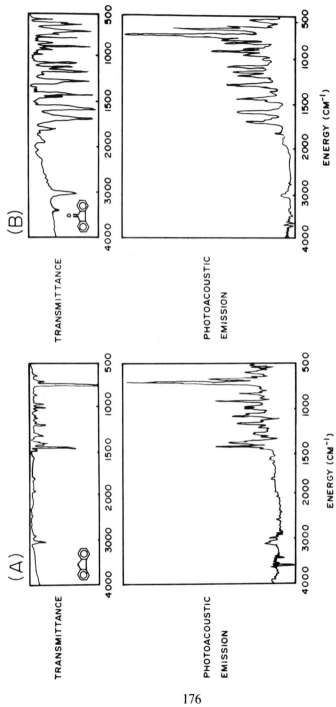

Figure 20. The IR transmission spectra (Sadtler index) compared to FT-IR-PAS emission spectra: (*a*) fluorene; (*b*) fluorenone. The PAS data are source compensated by deconvolution of the source envelope determined using a DTGS detector (G. E. Cabaniss and R. W. Linton, unpublished results, 1981).

176

within a factor of two of the true concentrations. As previously noted, XPS and AES are usually limited to surface elements present at about 1% or greater atomic concentration. Although a major advantage of PAS is its utility for the qualitative spectroscopic characterization of highly light scattering particulate samples, the quantitative relationships are not well established.

The general quantitative expression for PAS is quite complex (10). However, under idealized conditions for an optically opaque ($\mu_\beta \ll l$), and thermally thick ($\mu_s \ll l$; $\mu_s < \mu_\beta$) solid, the photoacoustic signal is proportional to $\eta\beta\mu_s$ [where l = length of solid perpendicular to excitation beam (cm); μ_s = thermal diffusion length (cm); μ_β = optical penetration depth (cm); β = optical absorption coefficient (cm^{-1}); η = fraction of absorbed energy that is involved in nonradiative relaxation].

The β term is the product of the concentration and the absorptivity of the analyte. Some efforts have been made to use phase information to improve the quantitative relationships of PAS (146, 147), but the phase behavior of particulate materials has yet to be clearly delineated (146). Leyden and co-workers (147) have made an attempt to correct for scattering in highly reflective particulate materials. This scattering may increase photoacoustic signals owing to increased surface area irradiation. The reflectivity of adsorbents decreases, however, if a highly radiation absorbing material is deposited on the surface. Sample dilution with uncoated reflective material also changes the thermal diffusion characteristics of the filler gas at the very critical solid–gas interface (148). Cooler gas in the area surrounding the uncoated reflective surfaces facilitates thermal diffusion from the coated particles causing enhanced photoacoustic intensity. For example, the photoacoustic intensity of mixtures of 4 mg/g fluorene adsorbed on silica and diluted 1:1, 1:3, and 1:7 with uncoated silica shows only slight variation (149).

Because XPS and AES elemental sensitivities show limited variations with atomic number (Figure 9) or with changes in sample matrix composition, an empirical elemental sensitivity factors approach can be used routinely to provide semiquantitative surface compositions with typical uncertainties of 20–30%. For example, in XPS relative elemental concentrations can be determined with atomic sensitivity factors (ASF):

$$\frac{n_A}{n_B} = \frac{I_A/S_A}{I_B/S_B} \tag{5}$$

where n is the atomic concentration of A or B, I is the intensity (area) of the XPS peak and S is the atomic sensitivity factor. The XPS sensitivity factor for a given element can be derived from different samples containing the element of interest referenced to an arbitrarily chosen elemental line (e.g., fluorine ($1s$), ASF = 1.0). An analogous approach is used in Auger electron spectroscopy quantitation. Tabulated sensitivity factors are available in handbook form (150), but are suspect, in part owing to apparent variations in the response functions for different XPS or Auger electron spectroscopy instruments (151). Another limitation in both XPS and AES is the lack of hydrogen information. Hydrogen often is present in major surface atomic concentrations, for example on particulates having substantial amounts of sorbed water or organic materials. Purely empirical quantitation using particulate standards also is of limited utility owing to the general physical and chemical complexity of environmental particles and the difficulty of generating standard particles with surface compositions reflective of bulk compositions (47).

Fundamental or "first principles" approaches to XPS or Auger electron spectroscopy quantitation also exist. For example, in XPS the detected signal intensity as indicated by its peak area can be related to the surface elemental concentration:

$$I = F\sigma\lambda n f_{\text{geom}} b_{\text{det}} C \qquad (6)$$

where F is the x-ray flux, σ is the photoionization cross section for a particular core energy level, λ is the photoelectron mean free path, f_{geom} is the geometric and transmission factor for the spectrometer, b_{det} is the detector sensitivity, C is the fractional efficiency for photoelectron emergence through a contaminating adsorbed layer, and other terms (I, n) are as previously defined. For most cases F, C, and ($f_{\text{geom}} \times b_{\text{det}}$) are assumed constant, thus relative concentrations can be calculated:

$$\frac{n_A}{n_B} = \frac{I_A/\sigma_A\lambda_A}{I_B/\sigma_B\lambda_B} \qquad (7)$$

Photoionization cross sections have been published by Scofield (152) and λ values can be calculated or approximated using empirical compilations (122). More details on both of these methods can be found elsewhere (153). Comparable "first principles" expressions are available in electron-excited Auger electron spectra that contain additional parameters for the Auger electron yield (x-ray emission is a competitive process) and a back-scattering factor that corrects for core-level ionization enhancements due to backscattered primary electrons (154). Limited information on the sample matrix dependence of the Auger yields and backscattering factors are barriers with respect to the application of the "first principles" Auger electron spectroscopy approach to the determination of surface elemental concentrations.

All the preceding quantitative methods for XPS and Auger electron spectroscopy assume a sample that is homogeneous both laterally and in depth. Alternative methods must be used if accurate quantitative information is to be obtained for typical environmental particulate samples consisting of a bulk substrate with adsorbed or chemically unique surface layers. Approximately 0.1 at. % of an element can be detected in bulk samples (30) and as little as 0.01 of an adsorbed monolayer in ideal situations (155).

Equations for Auger electron spectroscopy or XPS intensity can be derived that take into account an exponential attenuation of the Auger or x-ray photoelectrons passing through an adsorbed layer. For example, the XPS intensity for a bulk element, I_B, is described by

$$I_B = \frac{K\sigma_B\chi_B g\lambda_B}{E_B} \exp\left(\frac{-z}{g\lambda_B}\right) \qquad (8)$$

where K is a constant accounting for instrumental factors, χ_B is the number of atoms per unit volume, g is a specimen-spectrometer geometrical factor, E_B is the photoelectron kinetic energy, z is the adsorbed layer thickness, and σ and λ are as previously defined. The XPS intensity for an element in the adsorbed layer is then:

$$I_A = \frac{K\sigma_A\chi_A g\lambda_A}{E_A} \left[1 - \exp\left(\frac{-z}{g\lambda_A}\right) \right] \qquad (9)$$

From these equations, surface thickness and elemental concentrations can be derived. Carbon surface contamination can present a problem since all photoelectrons from the sample must pass through the contaminating layer. Dreiling (156) has published a more extensive discussion of these equations and their applications.

Despite the larger variation in SIMS elemental sensitivities as a function of atomic number (in comparison to XPS and Auger electron spectroscopy), an analogous empirical sensitivity factors approach again may be used (47):

$$\frac{n_A}{n_B} = \frac{I_A/S_A f_A}{I_B/S_B f_B} \tag{10}$$

where the I's are the measured secondary ion intensities, S's are the relative elemental sensitivity factors derived from known standards, f's are the isotopic abundances of the lines measured, and the n's are the atomic concentrations of elements A and B. Although absolute sensitivities vary strongly with the sample matrix composition, relative sensitivities are less susceptible to matrix effects (41, 47). This is especially true for the highly oxidized matrices typical of many environmental particles. Analysis using a primary oxygen beam further increases the oxygen content of the surface and minimizes the effect of different electropositive species in the sample matrix. For example, Newbury and co-workers have shown that 80% of SIMS analyses performed on multicomponent silicate glasses fall within a factor of 2 of the "true" concentrations when elemental sensitivity factors are used (47).

Although several fundamental models are available to predict secondary ion emission (41, 157), none show substantial quantitative improvement over the sensitivity factors approach. Although the local thermal equilibrium (LTE) model has been the most successful for highly oxidized matrices, its application to the multicomponent glasses described earlier did not compare favorably with the ASF approach (47). Specifically, 80% of the LTE analyses fell only within a factor of 5 of the true concentrations.

4.5. Sample Preparation and Vacuum Requirements

Of the surface techniques reviewed, only PAS can be performed on particulate samples without the necessity of high vacuums in the analytical chamber. This is a particular advantage in the characterization of volatile sorbed species. For example, results of XPS studies of nitropyrene sorbed on coal fly ash are shown in Figure 21 (149). At ambient temperatures, fly ash constituents (Fe, Al, Si, O, S) are observed, but nitrogen is not suggesting desorption of the volatile nitropyrene. Experiments using a sample probe cooled to $-150°C$ are limited by water adsorption, which obscures even the fly ash matrix elements (Figure 21). The low atomic fraction (<4%) of nitrogen in nitropyrene also is a barrier to XPS detection, although very high surface coverages on fly ash have been characterized by XPS (158).

Indium foil is the most widely used substrate for mounting particulate samples for XPS, Auger electron spectroscopy, or SIMS characterization since particles are easily embedded and the foil is electrically conductive (78, 87). Indium foils also can be used directly as collection surfaces within cascade impactors or particle collection devices (86). A potential limitation of mounting small particles on a metal substrate is sputter deposition of the substrate material over the particles during depth-profiling experiments (87).

Analyzing aquatic particles with XPS requires that the sample be transferred from an aqueous solution to an ultrahigh vacuum (UHV). Consideration must be given to how the sample will be removed from solution and mounted for analysis and to how the UHV environment will affect the sample.

Sample mounting techniques that have been used for XPS include air

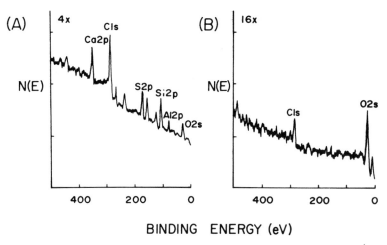

Figure 21. XPS data for coal fly ash with sorbed l-nitropyrene: (a) ambient ultrahigh vacuum conditions (10^{-9} Torr); (b) low-temperature experiment ($-150°C$) using liquid nitrogen cooled sample probe. (G. E. Cabaniss and R. W. Linton, unpublished results, 1981).

drying of the particles and mounting on double-sided sticky tape (57), suspending the particles in acetone or water by sonification and depositing as a slurry directly on the sample probe (63), depositing the particles as an aqueous slurry on indium foil or acid-etched aluminum foil (58), and drying the particles in a vacuum oven and pressing into a Cu grid (71). The deposition of hydrous ferric oxide particles on a liquid nitrogen cooled cold probe was found to be of limited utility due to the formation of a surface layer of ice (58). Sample mounting was facilitated in some studies by using cleaved crystal surfaces (69) or thin films (67) as the adsorbing surface.

Contamination from the residual fraction of the adsorbing species presents a potential problem. In most aqueous adsorption studies, the samples are rinsed prior to analysis to remove any nonadsorbed ions. Even when samples have not been rinsed, however, contamination appears to be minimal (Figure 14).

One potentially serious problem in the XPS analysis of aqueous adsorption is the effect of transferring a sample from an aqueous to a UHV environment. If XPS results are to have validity, then the degree of surface alteration due to the desorption of physically or chemically adsorbed water must be evaluated. Harvey and Linton (58) determined that the surface alteration of HFO was apparently limited to the desorption of physically adsorbed water. The underlying chemical composition appeared to survive largely intact on transferring the HFO from solution to the UHV environment of the spectrometer.

REFERENCES

1. E. Denoyer, R. Van Grieken, F. Adams, and D. F. S. Natusch, *Anal. Chem.* **54**, 26A (1982).

2. J. J. Blaha, E. S. Etz, and K. F. J. Heinrich, Raman Microprobe Analysis of Stationary Source Particulate Pollutants, U.S. Environmental Protection Agency Report, No. 600/2-80-173, 32 pp., 1980.

3. J. A. Small, K. F. J. Heinrich, D. E. Newbury, R. L. Myklebust, and C. E. Fiori, Procedure in the Quantitative Analysis of Single Particle with the Electron Probe, in *Characterization of Particles*, K. F. J. Heinrich, Ed., National Bureau of Standards Special Publication 533, U.S. Government Printing Office, Washington, D.C., 1980, p. 29.

4. A. G. Bell, *Am. J. Sci.* **20**, 305 (1880).

5. M. L. Viengerov, *Dokl. Akad. Nauk. SSSR* **19**, 687 (1938).

6. A. H. Pfund, *Science* **90**, 326 (1939).

7. J. G. Parker, *Appl. Opt.* **12**, 2974 (1973).

8. A. Rosencwaig, *Photoacoustics and Photoacoustic Spectroscopy*, Wiley-Interscience, New York, 1980.

9. A. Rosencwaig and A. Gersho, *Science* **190**, 556 (1975).

10. A. Rosencwaig and A. Gersho, *J. Appl. Phys.* **47**, 64 (1976).

11. *Instruction Manual, Model 6003 PAS Sample Cell and Model 6005 Preamplifier*, Princeton Applied Research, Princeton, New York, 1981.

12. M. J. D. Low and G. A. Parodi, *Spectrosc. Lett.* **13**, 151 (1980).

13. M. J. D. Low and G. A. Parodi, *Infrared Phys.* **20**, 333 (1980).

14. M. J. D. Low and G. A. Parodi, *Appl. Spectrosc.* **34**, 76 (1980).

15. M. J. D. Low and G. A. Parodi, *Spectrosc. Lett.* **13**, 663 (1980).

16. W. J. Boyd et al., *Rev. Sci. Instrum.* **45**, 1286 (1974).

17. D. W. Vidrine, *Appl. Spectrosc.* **34**, 314 (1980).

18. M. G. Rockley, *Appl. Spectrosc.* **34**, 405 (1980).

19. M. G. Rockley and J. P. Devlin, *Appl. Spectrosc.* **34**, 407 (1980).

20. M. G. Rockley, *Science* **210**, 918 (1980).

21. M. G. Rockley et al., *Appl. Spectrosc.* **35**, 185 (1981).

22. G. Laufer et al., *Appl. Phys. Lett.* **37**, 517 (1980).

23. N. Teramae et al., *Chem. Lett.*, 1091 (1981).

24. S. M. Riseman and E. M. Eyring, *Spectrosc. Lett.* **14**, 163 (1981).

25. J. B. Kinney et al., *J. Am. Chem. Soc.* **103**, 4273 (1981).

26. D. G. Mead et al., *Int. J. Infrared Millimeter Waves* **2**, 23 (1981).

27. S. Perkowitz and G. Busse, *Opt. Lett.* **5**, 228 (1980).

28. P. R. Griffths, *Infrared Fourier Transform Spectroscopy*, Wiley-Interscience, New York, 1975.

29. D. K. Killinger, et al., in *Laser Probes for Combustion Chemistry*, D. R. Crosley, Ed., ACS Symposium Series Volume 134, American Chemical Society, Washington, D.C., 1980, p. 457.

30. W. M. Riggs and M. J. Parker, in *Methods of Surface Analysis*, A. W. Czanderna, Ed., Elsevier Scientific, New York, 1975.

31. C. S. Fadley, in *Electron Spectroscopy—Theory, Techniques, Applications*, Volume 2, C. R. Brundle and A. D. Baker, Eds., Academic, New York, 1978.

32. D. Briggs, *Handbook of X-ray and UV Photoelectron Spectroscopy*, Heyden, Philadelphia, 1978.

33. T. Novakov, S.-G. Chang, R. L. Dod, and L. Gundel, Chapter 5 of this book.

34. A. Joshi, L. E. Davis, and P. W. Palmberg, in *Methods of Surface Analysis*, A. W. Czanderna, Ed., Elsevier Scientific, New York, 1975, Chapter 5.

35. J. Kirschner, in *Electron Spectroscopy for Surface Analysis*, H. Ibach, Ed., Springer-Verlag, New York, 1977, Chapter 3.

36. R. Vanselow and S. Y. Tong, *Chemistry and Physics of Solid Surfaces*, Chemical Rubber, Cleveland, 1977.

37. T. A. Carlson, *Photoelectron and Auger Spectroscopy*, Plenum, New York, 1975.

38. P. W. Palmberg, *J. Electron Spec. Rel. Phenom.* **5**, 691, (1974).

39. P. W. Palmberg and W. M. Riggs, "A Unique Instrument for Multiple Technique Surface Characterization by ESCA, Scanning Auger, UPS, and SIMS", Proceedings of the Seventh International Vacuum Congress and Third International Conference on Solid Surfaces, Vienna, 1977.

40. J. P. Thomas and A Cachard, Eds., *Materials Characterization Using Ion Beams*, Plenum, New York, 1978.

41. K. F. J. Heinrich and D. E. Newbury, Eds., *Secondary Ion Mass Spectrometry*, National Bureau of Standards Special Publication 427, Washington, D.C., 1975.

42. *Scanning* **3** (2), (1980). (Entire issue devoted to SIMS.)

43. A. Benninghoven, C. A. Evans, Jr., R. A. Powell, R. Shimizu, and H. A. Storms, Eds., *Secondary Ion Mass Spectrometry—SIMS II*, Springer-Verlag, New York, 1979.

44. H. Liebl, in Reference 41, Chapter 1.

45. G. H. Morrison and G. Slodzian, *Ann. Chem.* **47** (11), 932A (1975).

46. R. J. Blattner, C. A. Evans, Jr., *Scanning Electron Microsc.* **4**, 55 (1980).

47. D. E. Newbury, in *Characterization of Particles*, K. J. J. Heinrich, Ed., National Bureau of Standards Special Publication 533, Washington, D.C., 1980, Chapter 11.

48. B. K. Furman and G. H. Morrison, *Anal. Chem.* **52**, 2305 (1980).

49. K. K. Killinger and S. M. Jaspar, *Chem. Phys. Lett* **66**, 207 (1979).

50. D. M. Roessler and F. R. Foxvog, *Appl. Opt.* **18**, 1399 (1979).

51. G. E. Cabaniss and R. W. Linton, in *Polynuclear Aromatic Hydrocarbons—Fifth International Symposium*, M. Cooke and A. J. Dennis, Eds., Battelle Press, Columbus, OH, 1981, p. 277.

52. A. H. Miguel, et al., *Environ. Sci. Technol.* **13**, 1229 (1979).

53. J. L. Howell and R. A. Palmer, "Photoacoustic Analysis of Atmospheric Aerosol Filters," Pittsburgh Conference on Analytical Chemistry and Applied Spectroscopy, Atlantic City, NY, 1980, Paper No. 675.

54. J. P. Monchalin et al., *Appl. Phys. Lett.* **35** (5), 360 (1979).

55. K. K. Krauskoph, *Geochim. Cosmochim. Acta* **9**, 1 (1956).

56. E. A. Jenne, *Adv. Chem. Ser.* **73**, 337 (1968).

57. M. H. Koppelman and J. G. Dillard in *ACS Symp. Ser.* **18**, 186 (1975).

58. D. T. Harvey and R. W. Linton, *Anal. Chem.* **53**, 1689 (1981).

59. W. Stumm and J. J. Morgan, *Aquatic Chemistry*, Wiley-Interscience, New York, 1970, p. 527.

60. M. Kabayashi and M. Uda, *J. Non-Cryst. Solids* **29**, 419 (1978).

61. M. E. Counts, J. S. C. Jen, and J. P. Wightman, *J. Phys. Chem.* **77**, 1924 (1973).

62. J. M. Adams and S. Evans, *Clays Clay Miner.* **27**, 248 (1979).

63. M. H. Koppelman, A. B. Emerson, and J. G. Dillard, *Clays Clay Miner.* **28**, 119 (1980).

64. M. H. Koppelman and J. G. Dillard, *Clays Clay Miner.* **25**, 457 (1977).

65. M. H. Koppelman and J. G. Dillard, *J. Colloid Interface Sci.* **66**, 345 (1978).

66. J. W. Murray and J. G. Dillard, *Geochim. Cosmochim. Acta* **43**, 781 (1979).

67. D. G. Brûlé, J. R. Brown, G. M. Bancroft, and S. W. Fyfe, *Chem. Geol.* **25**, 227 (1979).

68. K. K. Turekian and L. H. Wedepohl, *Bull. Geol. Soc. Am.* **72**, 175 (1961).

69. G. M. Bancroft, J. R. Brown, and W. S. Fyfe, *Chem. Geol.* **19**, 131 (1977).

70. J. R. Brown, G. M. Bancroft, W. S. Fyfe, and R. A. N. McLean, *Environ. Sci. Technol.* **13**, 1142 (1979).

71. P. H. Tewari and W. Lee, *J. Colloid Interface Sci.* **52**, 77 (1975).

72. R. Alvarez, C. S. Fadley, J. A. Silva, and G. Vehara, *Soil Sci. Soc. Am. Proc.* **40**, 615 (1976).

73. M. H. Koppelman and J. G. Dillard, *Clays Clay Miner.* **28**, 211 (1980).

74. C. J. Powell, in *Characterization of Particles*, K. J. J. Heinrich, Ed., National Bureau of Standards Special Publication 533, Washington, D.C., 1980, Chapter 10.

75. D. L. Davidson and E. M. Gause in *Characterization of Particles*, K. J. J. Heinrich, Ed., National Bureau of Standards Special Publication 533, Washington, D.C., 1980, Chapter 8.

76. T. R. Keyser, D. F. S. Natusch, C. A. Evans, Jr., and R. W. Linton, *Environ. Sci. Technol.* **12**, 768 (1978).

77. R. W. Linton, A. Loh, D. F. S. Natusch, C. A. Evans, Jr., and P. Williams, *Science* **191**, 852 (1976).

78. R. W. Linton, P. Williams, C. A. Evans, Jr., and D. F. S. Natusch, *Anal. Chem.* **49**, 1514 (1977).

79. P. A. Lindfors and C. T. Hovland, in *Environmental Pollutants—Detection and Measurement*, T. Y. Toribara, J. R. Coleman, B. E. Dahneke, and I. Feldman, Eds., Plenum, New York, 1978, p. 349.

80. R. W. Linton, in *Monitoring Toxic Substances*, D. Schuetzle, Ed., ACS Symposium Series, Volume 94, American Chemical Society, Washington, D.C., 1979, p. 137.

81. J. A. McHugh and J. F. Stevens, *Anal. Chem.* **44,** 2187 (1972).

82. D. F. S. Natusch, C. F. Bauer, H. Matusiewicz, C. A. Evans, Jr., J. Baker, A. Loh, R. W. Linton, and P. K. Hopke, *Proceedings of the International Conference on Heavy Metals in the Environment*, Volume 2, Part 2, Toronto, Canada, 1977, p. 553.

83. W. R. Roy, R. G. Thiery, R. M. Schuller, and J. J. Suloway, *Coal Fly Ash: A Review of the Literature and Proposed Classification System with Emphasis on Environmental Impacts*, Environmental Geology Note 96, Illinoise State Geological Survey, Champaign, IL April, 1981.

84. R. A. Powell and W. E. Spicer, *Characterization of Fly Ash and Related Oxides Using Auger Electron Spectroscopy*, Electric Power Research Institute (EPRI) Report FP-708, Palo Alto, CA, 1978.

85. J. L. Hock, D. Snider, J. Kovacich, and D. Lichtman, Electron Beam Effects in AES Studies of Micron Size Insulating Particles, University of Wisconsin–Milwaukee, submitted for publication, *Applic. Surf. Sci.,* 1981.

86. J. L. Hock and D. Lichtman, Studies of Surface Layers on Single Particles of Fly Ash, University of Wisconsin–Milwaukee, submitted for publication, *Environ. Sci. Technol.,* 1981.

87. J. L. Hock, D. Snider, R. Ford, and D. Lichtman, The Cover-up of Particles by Substrate Material During Ion Beam Sputtering, University of Wisconsin–Milwaukee, submitted for publication, *J. Vac. Sci. Technol.,* 1981.

88. J. A. King, H. Grimm, Entrained Cenospheres from a Fluidized-Bed Combustor, Morgantown Energy Technology Center, Morgantown, WV, submitted for publication, *Science,* 1981.

89. J. H. Faber, Fifth International Ash Utilization Symposium, METC/SP-79/10, 1979, p. 24.

90. G. L. Fisher, C. E. Chrisp, and W. G. Jennings, *Trace Elements Environ. Health* **12,** 293 (1978).

91. R. L. Davison, D. F. S. Natusch, J. R. Wallace, and C. A. Evans, Jr., *Environ. Sci. Technol.* **8** (13), 1107 (1974).

92. J. W. Kaakinen, R. M. Jordan, N. H. Lawasani, and R. E. West, *Environ. Sci. Technol.* **9,** 862 (1975).

93. D. H. Klein et al., *Environ. Sci. Technol.* **9,** 973 (1975).

94. C. B. Block and R. Dams, *Environ. Sci. Technol.* **10,** 1011 (1976).

95. D. G. Coles, R. C. Ragaini, J. M. Ondov, G. L. Fisher, D. Silbermann, and B. A. Prentice, *Environ. Sci. Technol.* **13,** 455 (1979).

96. G. L. Fisher, C. E. Chrisp, and O. G. Raabe, *Science* **204,** 879 (1979).

97. R. D. Smith, J. A. Campbell, and K. K. Nielson, *Environ. Sci. Technol.* **13,** 553 (1979).

98. D. F. S. Natusch and J. R. Wallace, *Science* **186,** 695 (1974).

99. D. F. S. Natusch, J. R. Wallace, and C. A. Evans, Jr., *Science* **183,** 202 (1974).

100. D. F. S. Natusch, *Environ. Health Perspectives* **22**, 79 (1978).

101. L. D. Hulett, H. W. Dunn, J. M. Dale, J. F. Emery, W. S. Lyon, and P. S. Murty, *Measurement Detection and Control of Environmental Pollutants—Proceedings of a Symposium*, Vienna, 1976, p. 29.

102. J. A. Campbell, R. D. Smith, and L. E. Davis, *Appl. Spectrosc.* **32**, 316 (1978).

103. Y. Mamane and R. F. Pueschel, *Geophys. Res. Lett.* **6** (2), 109 (1979).

104. D. J. Swaine, *Trace Elements Environmental Health* **11**, 107 (1977).

105. G. L. Fisher, D. P. Y. Chang, and M. Brummer, *Science* **197**, 553 (1976).

106. R. W. Linton, S. R. Walker, C. R. DeVries, J. D. Shelburne, and P. Ingram, *Scanning Electron Microsc.* **2**, 583 (1980).

107. C. R. DeVries, P. Ingram, W. F. Gutnecht, S. R. Walker, J. D. Shelburne, and R. W. Linton, *Lab. Invest.* **42**, 111 (1980).

108. R. W. Linton, D. F. S. Natusch, R. L. Solomon, and C. A. Evans, Jr., *Environ. Sci. Technol.* **14**, 159 (1980).

109. R. W. Linton, Ph.D. thesis, University of Illinois, Urbana, IL, 1977.

110. K. W. Olson and R. K. Skogerboe, *Environ. Sci. Technol.* **9**, 227 (1975).

111. M. E. Farmer and R. W. Linton, Federation of Analytical Chemistry and Spectroscopy Societies—Eighth Annual Meeting, Abstract No. 36, Philadelphia, PA, September, 1981.

112. F. E. Lundin, Jr., J. K. Wagoner, and V. E. Archer, NIOSH-NIEHS Joint Monograph No. 1, 1971.

113. E. A. Martrell, *Am. Sci.* **63**, 404 (1979).

114. C. A. Evans, Jr., *Anal. Chem.* **47** (9), 818A (1975).

115. H. A. Storms, K. F. Brown, and J. D. Stein, *Anal. Chem.* **49**, 2023 (1977).

116. C. G. Pantano and T. E. Madey, *Appl. Surf. Sci.* **7**, 115 (1981).

117. M. Lepareur, "Le microanalyseur ionique de seconde génération Cameca, Modéle 3F," in *Rev. Tech. Thomson CSF* **12** (1), 225 (1980).

118. V. L. Roggli, P. Ingram, R. W. Linton, W. F. Gutnecht, P. Mastin, and J. D. Shelburne, in *Pulmonary Toxicology*, G. E. R. Hook, Ed., Raven Press, New York, in press, 1982.

119. J. J. Cazaux, *Microsc. Spectrosc. Electron.* **1**, 73 (1976).

120. C. T. Hovland, *Appl. Phys. Lett.* **30**, 274 (1977).

121. J. A. Noonan and D. M. Munroe, *Opt. Spec.* **13** (2), 28 (1979).

122. M. S. Seah and W. A. Dench, *Surf. Interface Anal.* **1**, 2 (1979).

123. M. H. Koppelman, in *Advanced Chemical Methods for Soil and Clay Minerals Research*, J. W. Stucki and W. L. Banwart, Eds., D. Reidel, Dordrecht, The Netherlands, 1980, pp. 205–243.

124. J. W. Coburn, *J. Vac. Sci. Technol.* **13**, 1037 (1976).

125. J. S. Solomon and V. Meyers, *Am. Lab.* **8** (3), 31 (1976).

126. Y. M. Cross and J. Dewing, *Surf. Interface Anal.* **1** (1), 26 (1979).

127. M. J. Adams and G. F. Kirkbright, *Analyst* **102**, 281 (1977).

128. P. Helander et al., *J. Appl. Phys.* **52**, 1146 (1981).

129. G. Loforth, *Chemosphere* **7**, 791 (1978).

130. A. H. Miguel, Ph.D. thesis, University of Illinois, Urbana, IL, 1976.

131. W. A. Korfmacher, Ph.D. thesis, University of Illinois, Urbana, IL, 1978.

132. D. F. S. Natusch, *Environ. Health Perspect.* **22**, 78 (1978).

133. D. F. S. Natusch, C. F. Bauer, and A. Loh, in *Pollution Control*, Vol. III, W. Strauss, Ed., Wiley-Interscience, New York, 1978.

134. C. D. Wagner, L. H. Gale, and R. H. Raymond, *Anal. Chem.* **51**, 466 (1979).

135. D. T. Harvey and R. W. Linton, Eleventh Annual Symposium on the Analytical Chemistry of Pollutants, Jekyll Island, GA, May 1981.

136. D. F. Smith, Ph.D. thesis, University of North Carolina, Chapel Hill, NC, 1978.

137. G. C. Allen, M. T. Curtis, A. J. Hooper, and P. M. Tucker, *J. Chem. Soc. Dalton Trans.* **14**, 1525 (1975).

138. N. S. McIntryre and D. G. Zetaruk, *Anal. Chem.* **49**, 1521 (1977).

139. C. R. Brundle, T. J. Chuang, and K. Wandelt, *Surf. Sci.* **68**, 459 (1977).

140. C. D. Wagner, D. A. Zatko, and R. H. Raymond, *Anal. Chem.* **52**, 1445 (1980).

141. D. Lichtman, J. H. Craig, V. Sailer, and M. Drinkwine, *Appl. Surf. Sci.* **7**, 325 (1981).

142. H. H. Madden, *J. Vac. Sci. Technol.* **18** (3), 677 (1981).

143. R. J. Day, S. E. Unger, and R. G. Cooks, *Anal. Chem.* **52**, 557A (1980).

144. J. A. Gardella, Jr. and D. M. Hercules, Ninth Annual Symposium on the Analytical Chemistry of Pollutants, Jekyll Island, GA, May 1979.

145. U. R. Stenberg and T. E. Alsberg, *Anal. Chem.* **53**, 2067 (1981).

146. A. Mandelis et al., *J. Appl. Phys.* **50**, 7138 (1979).

147. L. W. Burggraf and D. E. Leyden, *Anal. Chem.* **53**, 2037 (1981).

148. J. K. Becconsall et al., *Anal. Chem.* **53**, 2037 (1981).

149. G. E. Cabaniss and R. W. Linton, unpublished results, 1981.

150. C. D. Wagner et al., *Handbook of X-ray Photoelectron Spectroscopy;* L. E. Davis, et al., *Handbook of Auger Electron Spectroscopy,* Physical Electronics–Perkin Elmer, Eden Prairie, MN, 1978.

151. C. J. Powell, *Appl. Surf. Sci.* **4**, 492 (1980).

152. J. H. Scofield, Lawrence Livermore Laboratory Report UCRL-51326, 1973.

153. C. J. Powell, in Reference 154, 1978, pp. 5–30.

154. N. S. McIntyre, Ed., *Quantitative Surface Analysis of Materials*, ASTM–STP 643, American Society for Testing of Materials, Philadelphia, PA, 1978.

155. G. M. Bancroft, J. R. Brown, and W. S. Fyfe, *Chem. Geol.* **25,** 227 (1979).

156. M. J. Dreiling, *Surf. Sci.* **71,** 231 (1978).

157. P. Williams, *Surf. Sci.* **90,** 588 (1979).

158. M. M. Hughes et al., in *Polynuclear Aromatic Hydrocarbons—Fourth International Symposium,* A. Bjorseth and A. J. Dennis, Eds., Battelle Press, Columbus, OH, 1979, p. 1.

ESCA IN ENVIRONMENTAL CHEMISTRY

T. NOVAKOV, S.-G. CHANG, R. L. DOD, and L. GUNDEL

Lawrence Berkeley Laboratory
University of California
Berkeley, California 94720

1. INTRODUCTION

Suspended particulate matter plays a major role in the overall air pollution problem. It is directly responsible for reduction of visibility and contributes to acidification of precipitation; when deposited in the lungs, it may cause adverse health effects. These particles can consist of solid or liquid substances. Sulfur, nitrogen, and carbon compounds are the major species from anthropogenic sources. In many circumstances these compounds may constitute about 80% of the dry mass of particulates resulting from fossil fuel combustion.

The results of traditional wet chemical analyses of aerosol particles suggest that the principal elements such as sulfur and nitrogen occur as stoichiometric inorganic compounds: ammonium sulfate, ammonium bisulfate, a mixture of the two, sulfuric acid, and ammonium nitrate. However, applications of physical methods for chemical characterization of aerosol particles such as electron spectroscopy for chemical analysis (ESCA) and electron diffraction often yield results that are not consistent with the results of wet chemical analyses.

The speciation of atmospheric aerosol particles is an important task because many of their environmental effects will depend on their specific chemical and physical states. It is also important to determine the chemical compounds and species as they actually exist in aerosol form and not as they may appear in aqueous solutions, which is what wet chemical methods reveal.

Such chemical analyses of sulfur and nitrogen species report only ions in solution. These ions, however, may be originally water soluble (e.g., nitrate and ammonium from ammonium sulfate), or they may be ionic products of hydrolyzable species. Of course, insoluble species will not be detected by wet chemical techniques.

In contrast, physical methods such as photoelectron spectroscopy (ESCA) (1) analyze the entire sample content without sample preparation. In this respect ESCA seems to have an advantage over wet chemical methods. Its application has produced important results on the chemical composition of atmospheric particulate matter. These results can be summarized as follows.

Sulfur is found to be predominantly in a $+6$ oxidation state, that is, sulfate. Other chemical states of sulfur have also been observed, although these seldom approach sulfate concentrations (2). Generally, at high pollutant concentrations, sulfate is practically the only sulfur species present. Nitrogen can be present in an oxidized and a reduced form. The oxidized form has been identified as nitrate, while the reduced form consists of ammonium and a group of species consisting of particulate amines and amides, N_x (3). These N_x species were first discovered by ESCA and are easily distinguished from ammonium by a ~2-eV chemical shift of the $N(1s)$ peak. ESCA also provided the first indication that, besides relatively stable ammonium nitrate, there is a volatile nitrate fraction probably associated with nitric acid adsorbed on particles or on the filter material (4, 5). ESCA measurements on ambient size-segregated samples have demonstrated for the first time that particulate nitrate is preferentially associated with particles greater than 2 μm. ESCA has also provided evidence that in some samples there is a volatile ammonium species in addition to the ammonium ion associated with sulfate and nitrate.

ESCA is a surface technique, and the sample is exposed to vacuum and x-ray bombardment during analysis. Therefore, ESCA results may not be representative of the bulk composition; some volatile species may be lost because of the vacuum, and in principle the x-ray bombardment may cause chemical changes of some species. Because of these possible problems, it seemed desirable to employ a technique that will analyze the bulk properties of particles without chemical treatment or preseparation of the sample and to compare these results with ESCA results. Thermal analysis in the evolved gas analysis (EGA) mode is one such technique (6–9).

In this paper we will present the results of ESCA analyses of particulate samples, supplemented with EGA and infrared spectroscopy.

2. EXPERIMENTAL METHODS

X-ray photoelectron spectroscopy (XPS), also known as ESCA (1), is a physical method well suited to chemical characterization of environmental pollutants such as airborne particulate matter. This method involves the analysis of the kinetic energies of photoelectrons expelled from a sample (usually solid) irradiated with monoenergetic soft x-rays. The kinetic energy of a photoelectron E_{kin}, emitted from an electron subshell i, is given by $E_{kin} = h\nu - E_i$, where $h\nu$ is the x-ray photon energy and E_i is the binding energy of an electron in that subshell. For a known x-ray energy, the determination of the photoelectron kinetic energies provides a direct measurement of the electron binding energies. The binding energies are characteristic of each element, which enables the identification of elements in the sample. The intensity of photoelectrons originating from a subshell of an element is proportional to the concentration of atoms of that element in the active sample volume. The sample volume is determined by the electron escape depth and the physical size of the sample. This feature enables the method to be used for quantitative elemental analysis.

The strength of ESCA is in its ability to distinguish different chemical *states* rather than different elements. The electron binding energies are not absolutely constant but are modified by the valence electron distribution. The binding energy of an electron subshell in a given atom varies slightly when this atom is in different chemical environments. For example, the nitrogen (1s) binding energy in a nitrate ion, NO_3^-, is greater than the nitrogen binding energy in a ammonium ion, NH_4^+, by about 5 eV. These differences in electron binding energy are known as the chemical shift.

The origin of the chemical shift can be understood in terms of the shielding of the core electrons by the electrons in the valence shell. A change in the charge of the valence shell results in a change of the shielding, which affects the core electron binding energies. For example, if an atom is oxidized, it donates its valence electrons and thus becomes positively charged with respect to its neutral configuration. Some of the shielding contribution to the total potential is removed so that the binding energy of the core electron has increased. Conversely, the binding energies will show an opposite chemical shift for reduced species. The usefulness of ESCA for the analysis of samples of unknown chemical composition lies in the determination of the chemical shift and its interpretation. In practice, measurements of the chemical shifts are supplemented by the determination of relative photoelectron intensities, from which information about the apparent stoichiometry of species can be obtained.

The ESCA used in more recent experiments is a modified AEI ES-200 electron spectrometer that has been updated by the installation of a Surface Science Laboratories Model 239G position-sensitive photoelectron detector. The modifications also included replacement of all lens and analyzer power supplies, as well as changing to a modern microprocessor-based data system. Data collection with the modified spectrometer is approximately 10 times as rapid as with the original, thus substantially decreasing sample degradation during analysis.

In EGA the sample is heated at a predetermined rate in an oxidizing or neutral atmosphere. The evolved gases resulting from volatilization, decomposition, and combustion of the sample are monitored as a function of temperature by one or more gas-specific detectors. The carrier gas is usually oxygen or nitrogen. For analysis of carbonaceous materials, the gas detected in the oxygen mode is CO_2 (8). For analysis of nitrogenous species, we use oxygen as the carrier gas and detect total nitrogen oxides, NO_x (9).

A schematic representation of the EGA apparatus used in our analysis of aerosol particles is shown in Figure 1. The sample, collected on a prefired quartz filter, is placed in the quartz combustion tube so that its surface is perpendicular to the tube axis. The tube is supplied with purified oxygen, with excess oxygen escaping through an axial opening at the end of the tube.

The remainder of the oxygen (together with gases produced during

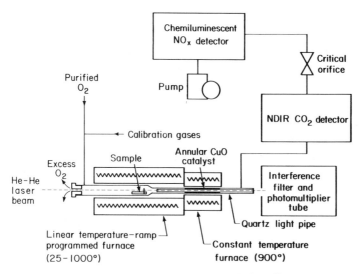

Figure 1. Schematic diagram of thermal analysis (EGA) apparatus.

analysis) is pulled at a constant rate determined by a critical orifice through a nondispersive infrared CO_2 analyzer (MSA LIRA 202S) and then through a chemiluminescent NO_x analyzer (Thermo-Electron Model 14D). Material may be evolved from the sample by volatilization, pyrolysis, oxidation, or decomposition. To ensure complete conversion of all carbon to CO_2, a section of the quartz tube immediately outside the programmed furnace is filled with a CuO catalyst bed, which is kept at a constant 900°C by a second furnace. This is especially necessary at relatively low temperatures (<250°C) where volatilization and incomplete combustion are the dominant processes occurring in the carbonaceous component.

Analyte gas concentrations are monitored as a function of temperature, and the resultant "thermogram" is a plot of concentration versus temperature with the integrated area of the curves being proportional to the carbon or nitrogen content of the sample. Quantitation is effected by calibration with gases of known concentration and by measuring the gas flow rate through the system. This calibration is verified by analyzing samples of quantitatively known elemental content.

The thermograms of ambient and source aerosol samples reveal distinct features in the form of peaks or groups of peaks. One important component of the carbonaceous aerosol is graphitic carbon, which is known to cause the black or gray coloration of ambient and source particulate samples (10). To determine which of the thermogram peaks corresponds to this graphitic carbon, we monitor the intensity of a He–Ne laser beam which passes through the filter. This provides simultaneous measurement of sample absorptivity and CO_2 evolution. The light penetrating the filter is collected by a quartz light guide and filtered by a narrow band interference filter to minimize the effect of the glow of the furnaces. An examination of the CO_2 and light intensity traces enables the assignment of the peak or peaks in the thermograms corresponding to the black carbon because they appear concurrently with the decrease in sample absorptivity.

The potential of this method (in the CO_2 mode) (9) is illustrated in Figure 2, where the complete thermogram of an ambient sample is shown. The lower trace represents the CO_2 concentration, while the upper curve corresponds to the light intensity of the laser light beam that reaches the detector during the temperature scan. Inspection of the thermogram shows that a sudden change in the light intensity occurs concomitantly with the evolution of the CO_2 peak at about 470°C. The light intensity I_0,

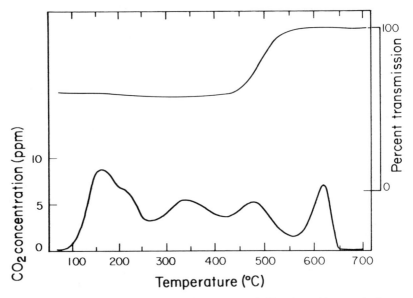

Figure 2. CO_2 and optical thermogram of a Berkeley, California, ambient particulate sample.

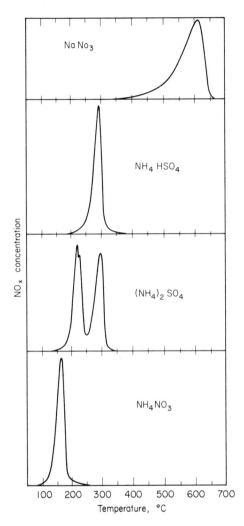

Figure 3. NO$_x$ EGA thermograms of nitrogen standard compounds.

after the 470°C peak has evolved, corresponds to that of a blank filter. This demonstrates that the light-absorbing species in the sample are combustible and carbonaceous—the graphitic carbon referred to earlier. The carbonate peak evolves at about 600°C; and as carbonate is not light absorbing, it does not change the optical density of the sample. In addition to black carbon and carbonate, the thermogram in Figure 2 shows several distinct groups of peaks at temperatures below ~400°C that correspond to various organics.

The potential for applications of EGA to the characterization of nitrogenous species is illustrated in Figure 3, where NO_x thermograms of NH_4NO_3, $(NH_4)_2SO_4$, NH_4HSO_4, and $NaNO_3$ are shown (11). It is obvious from the figure that distinguishing the principal nitrogen species is feasible by this technique.

For analysis by infrared spectroscopy, we used a Fourier-transform infrared spectrometer, consisting of an EOCOM model FMS 7001 P Michaelson interferometer and a PDP-11 computer.

3. RESULTS

3.1. Chemical States of Aerosol Sulfur and Nitrogen by ESCA and EGA

As an example of the capability of ESCA for differentiating different forms of atmospheric sulfates, Figure 4 shows the nitrogen ($1s$) and sulfur ($2p$) regions in ESCA spectra of two ambient samples, one from West Covina, California, and the other from St. Louis, Missouri. The peak positions corresponding to NH_4^+, N_x, and SO_4^{2-} are indicated. The solid vertical bar indicates the ammonium peak intensity expected under the assumption that the entire sulfate is in the form of ammonium sulfate. Obviously, the observed ammonium content in the West Covina sample is insufficient to account for the sulfate by itself. This is in sharp contrast to the St. Louis sample, where the observed ammonium intensity closely agrees with that expected for ammonium sulfate (4, 5).

These results demonstrate that ammonium sulfate in the aerosols can easily be distinguished from other forms of sulfate, such as the one found. However, wet chemical analyses (12) performed on West Covina samples collected simultaneously with the ESCA samples resulted in ammonium concentrations substantially higher than those suggested by the ESCA measurements. The discrepancy in ammonium determination by the two techniques is not caused by the volatilization of ammonium sulfate in the ESCA spectrometer vacuum, as evidenced by the St. Louis sample, where no volatile losses were observed. Ammonium bisulfate was also found to be stable in vacuum; and ammonium nitrate (negligible in these samples), while volatile in vacuum, is stable enough to be detected and determined in the time periods usually required to complete analysis.

Wet chemical analyses performed on other samples also contradict the

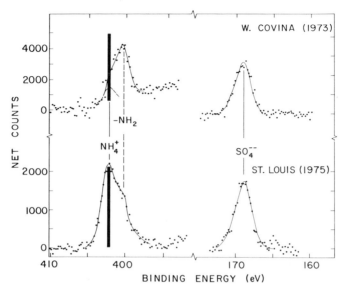

Figure 4. Nitrogen (1s) and sulfur (2p) regions in x-ray photoelectron spectra of two ambient samples. The peak positions corresponding to NH$_4{}^+$, —NH$_x$, and SO$_4{}^{2-}$ are indicated. The solid vertical bar represents the ammonium intensity expected under the assumption that the entire sulfate is in the form of ammonium sulfate. The difference in the relative ammonium content of the two samples is obvious. The sulfate and ammonium intensities in the St. Louis sample are compatible with ammonium sulfate. The ammonium in the West Covina sample is insufficient to be compatible with ammonium sulfate. Both samples were exposed to the spectrometer vacuum for about 1 h.

conclusions reached from ESCA studies (12). For example, total reduced nitrogen (NH$_4{}^+$ + N$_x$) as determined by ESCA often corresponds to the ammonium concentration determined by wet chemical methods. A consequence of this discrepancy is that in samples where wet analysis would indicate ammonium sulfate, ESCA would suggest an ammonium-deficient sulfate such as ammonium bisulfate, based on the assumption that the only nitrogen species that can act as a counterion for sulfate is ammonium. However, it has been demonstrated that a large fraction of the N$_x$ reduced nitrogen present in aerosol particles may be hydrolyzed to ammonium and removed by water extraction (13). Therefore, in contrast to ESCA, wet chemical analysis would indicate the presence of ammonium ion in sufficient concentration to account for the sulfate.

A tentative explanation of this observation is that the sample does not contain pure ammonium sulfate, but rather that some of the ammonium ions are replaced by, for example, a charged organic nitrogen complex. This complex should produce the right chemical shift for N_x (relative to ammonium) in the ESCA spectrum and hydrolyze to ammonium in water solution.

It is interesting to note that possibly related inconsistencies have also been observed by electron microscopy and electron diffraction (14). The principal conclusions from one such study are that sulfur, except mineral sulfate, is consistently associated with carbonaceous particles and that this sulfur is not in the form of ammonium sulfate, ammonium bisulfate, sulfamic acid, or sulfuric acid.

As we indicated earlier, it is commonly assumed that the only major particulate nitrogen species are ammonium and nitrate. However, from the very first attempts to apply ESCA for chemical analysis of atmospheric particles (1), it became clear that certain other reduced nitrogen species occur in concentrations similar to that of ammonium. ESCA measurements provided direct evidence for their existence, but ESCA alone could not tell what these species are—whether this is a single species or a group of compounds that exhibit a similar chemical ESCA shift.

Attempts made to prove the existence of the nonammonium particulate reduced nitrogen species by other techniques have met with only limited success. An attempt to detect N_x in ambient particles by infrared spectroscopy (15) failed because of interfering absorption bands. Combustion analysis provided the indirect evidence that nitrate and ammonium alone cannot account for total particulate nitrogen. However, the observed nitrogen deficiency was much less than indicated by ESCA.

In order to clarify this disagreement, we undertook a series of experiments (13) to study the solubility of particulate nitrogen species using ESCA for speciation and proton activation analysis (PAA) (16) to determine total nitrogen content. A series of ambient particulate samples collected on silver filters during autumn and winter in Berkeley, California, were sequentially washed with water and organic solvents, and analyzed by PAA and ESCA after each step. The ESCA results in Figure 5 show the effect of water washing on one of the samples (December 1977). The other samples exhibit similar behavior except that they contain less nitrate. The sulfate concentration in this and other samples is low. The nitrate peak occurs at binding energy of ~ 407 eV; NH_4^+, ~ 402 eV; and N_x, ~ 400 eV. The reduced nitrogen peaks have been deconvoluted into contributions from NH_4^+ and N_x by comparison with standards. The

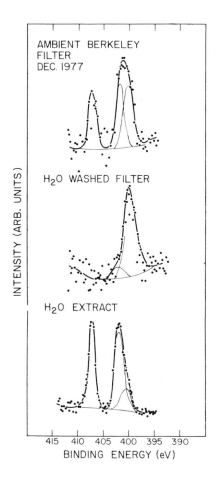

Figure 5. ESCA spectra of the nitrogen (1s) region of an ambient Berkeley, California, sample, H_2O washed and dried H_2O extract. The binding energies for NO_3^-, NH_4^+, and N_x are 407, 402, and 400 eV, respectively.

untreated particulate material contains roughly equivalent amounts of ammonium, N_x, and nitrate species, with total nitrogen equal to 61 $\mu g/cm^2$. After washing with water, nearly all the nitrogen on the filter is N_x. Ninety-six percent (59 $\mu g/cm^2$) of the initial nitrogen was removed by water extraction. The chemical speciation of the evaporated extract might be expected to sum with that of the extracted filter to yield the initial speciation; however, as seen from Figure 5, the extract contains much less N_x than NH_4^+ and shows a higher NO_3^- to NH_4^+ ratio.

It is not surprising that ESCA still detects N_x in the sample after water washing, which removed most of the nitrogen, since these surface groups would not be expected to be soluble. These findings suggest that much of the N_x present in the untreated sample has been converted to NH_4^+ during the extraction procedure, assuming that all nitrogen removed in water washing appears in the dried extract. If a substantial part of N_x were amides, their hydrolysis would yield NH_4^+. The expected reaction, illustrated for particulate amides, is

$$R—C—NH_2 + H_3O^+ \rightarrow R—C—OH + NH_4^+$$
$$\underset{O}{\overset{\|}{}} \qquad\qquad \underset{O}{\overset{\|}{}}$$

Removal of N_x by dissolution in water without hydrolysis can occur when N_x species are soluble stoichiometric compounds such as amines. Our results indicate that more than half the original N_x is removed by hydrolysis; some is removed by dissolution; and some remains on the filter, as expected for a mixture of amine, amide, and nitrile functional groups. Some of the species may be present as stoichiometric compounds and others possibly occur as surface functional groups on large carbon-aceous macromolecules.

These results clearly demonstrate that extraction of particulate material with water may chemically change the nitrogen species so that the chemical composition of the extract is not representative of the original sample. Therefore, analytical methods based on extraction may give erroneous results. The conversion of particulate amides to NH_4^+ during extraction will yield NH_4^+ concentrations that are too high.

Until quite recently the only direct proof for the existence of N_x was from ESCA measurements. It seemed to us that EGA in the NO_x mode was an ideal way to detect N_x. In order to identify N_x species by EGA, we need the results of our earlier ESCA analyses (17), which demonstrated that exposing an ambient sample to vacuum and x-ray bombardment for an extended time results in the volatilization of most nitrate and ammonium salts (additional heat is necessary to decompose ammonium sulfates). Thus, in a sample exposed to vacuum under these conditions, the only major nitrogenous species remaining will be N_x. This is illustrated in Figure 6, where the NO_x thermograms and ESCA spectra of the original and vacuum-exposed samples are shown (11). The ESCA spectra show that the principal species are volatile nitrate and ammonium and non-volatile N_x. Ammonium ions could be associated with either nitrate or sulfate, but the volatility indicated nitrate to be more probable. The cor-

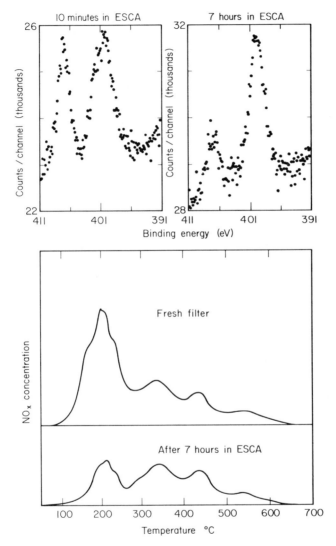

Figure 6. ESCA spectra and NO$_x$ thermograms from an ambient particulate sample from Riverside, California, before and after exposure to ESCA vacuum and x-rays. The ESCA spectrum of the fresh filter shows NO$_3^-$ (407 eV), NH$_4^+$ (402 eV), and N$_x$ (400 eV). The NO$_3^-$ and NH$_4^+$ components are largely missing after 7 h.

203

responding thermograms show a reduction in intensity in the triplet located between 150 and 300°C assigned to ammonium nitrate and ammonium sulfate. Peaks at ~350 and 450°C and the peak at ~550°C are unchanged, however. The first two probably correspond to N_x, while the last peak is most likely due to a metal nitrate.

In the example just described, the sulfate concentration is relatively small. As the ESCA spectrum in Figure 6 shows, nitrate could engage about 50% of the total reduced nitrogen as NH_4NO_3. Analysis of samples with different sulfate concentrations by EGA and ESCA has provided preliminary evidence that N_x functional groups may be directly associated with sulfate.

Figure 7 shows the ESCA spectrum of the N(1s) region of a sample

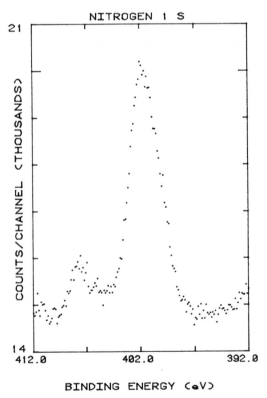

Figure 7. N(1s) ESCA spectrum of April 24, 1979 ambient particulate sample from Gaithersburg, Maryland. The dominant nitrogen species is NH_4^+ (402 eV), with small components of NO_3^- (407 eV) and N_x (400 eV).

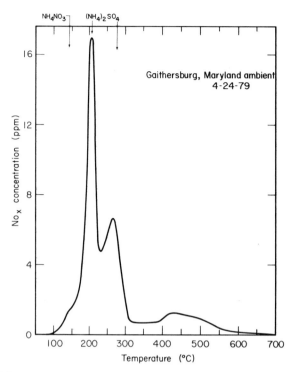

Figure 8. NO$_x$ thermogram of April 24, 1979 ambient particulate sample from Gaithersburg, Maryland.

collected in Gaithersburg, Maryland (11). In addition to a small nitrate peak, a pronounced ammonium peak is also seen. This sample contains a very small concentration of N$_x$ as evidenced only by a slight asymmetry of the ammonium peak. This sample contains enough ammonium to almost completely neutralize the sulfate as determined by ESCA. We conclude that here the sulfate is in the form of ammonium sulfate. The NO$_x$ thermogram of the same sample is shown in Figure 8. The positions of NH$_4$NO$_3$ and (NH$_4$)$_2$SO$_4$ thermogram peaks are indicated in the figure. This thermogram suggests that the principal counterion for sulfate is ammonium, not N$_x$, and that the sulfate is present as (NH$_4$)$_2$SO$_4$ (11). This is consistent with the ESCA observations.

The situation is quite different for the Anaheim sample whose N(1s) ESCA spectrum is shown in Figure 9. Here in addition to NO$_3$$^-$ and NH$_4$$^+$, the N$_x$ peak is clearly seen. Volatility in ESCA vacuum indicates

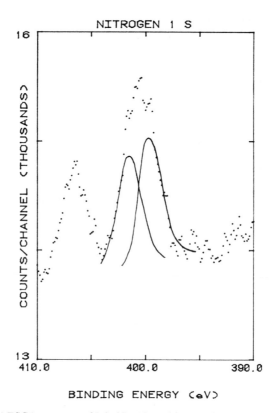

Figure 9.   N($1s$) ESCA spectrum of July 27, 1979 ambient particulate sample from Anaheim, California. The nitrogen is evenly divided among NO_3^- (407 eV), NH_4^+ (402 eV), and N_x (400 eV). The solid line shows deconvoluted NH_4^+ and N_x components.

that the nitrate in this sample is present as NH_4NO_3. Therefore, a significant portion of the ammonium is associated with the nitrate. The remainder of the ammonium is insufficient to provide counterions for the sulfate. This ESCA result suggests an ammonium-deficient sulfate compound such as ammonium bisulfate (11). We note that if the entire reduced nitrogen were hydrolyzed (as may occur during wet chemical analysis), there would be sufficient ammonium to conclude that the sulfate is present as ammonium sulfate.

The thermogram of this sample is shown in Figure 10. The ammonium nitrate peak and the first peak of ammonium sulfate match closely the positions of standard compounds. However, the next peak is noticeably shifted from the second ammonium sulfate peak, although the overall appearance of the doublet is similar to that of ammonium sulfate (11).

So far, in general we have found that the NO_x thermograms of ambient samples are consistent with $(NH_4)_2SO_4$ when the sulfur to carbon ratio is high and anomalous when the ratio is low. The observations just described cannot be fully explained at this time, but we can provide a tentative hypothesis that will be examined during the course of our work on this project. The doublet seen in the $(NH_4)_2SO_4$ thermogram is the result of the fact that ammonium sulfate thermally decomposes through NH_4HSO_4. The first peak at ~210°C is the result of the loss of NH_3 from $(NH_4)_2SO_4$. The second peak at ~270°C reflects the decomposition of the bisulfate into NH_3 and H_2SO_4. The two ammonium ions, of course, are indistinguishable by ESCA. An inspection of the anomalous NO_x ther-

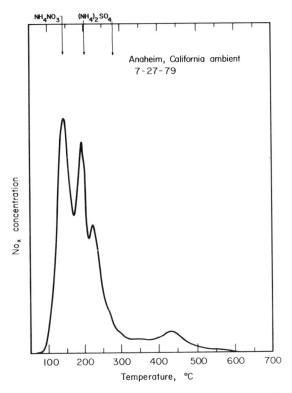

Figure 10. NO_x thermogram of July 27, 1979 ambient particulate sample from Anaheim, California.

mograms indicates that the high-temperature N_x peaks comprise a much smaller fraction of total nitrogen than indicated by ESCA (cf. Figure 9). Therefore a component of N_x could be associated with another peak such as the peak shifted to lower temperature from the second peak of ammonium sulfate. That shifted peak could represent a charged organic amide which serves in place of NH_4^+ as the second counterion for sulfate. Such species will produce an ESCA shift compatible with N_x, decompose at lower temperature than NH_4HSO_4, and hydrolyze to NH_4^+. Such a hypothesis could explain the observed inconsistencies between wet and other analytical methods. More importantly it would indicate that there is a direct chemical linkage of the carbonaceous fraction of the aerosol particles with sulfate and possibly nitrate.

3.2. Characterization of Reduced Aerosol Nitrogen by ESCA and Infrared Spectroscopy

Although the infrared spectroscopic technique has been used for the characterization of chemical species in ambient particulates (18, 19), reduced nitrogen species have not been revealed by this method because of their low infrared absorption cross sections and the fact that many chemical species in ambient particulates contribute to the absorption in the same infrared regions as the reduced nitrogen species. The characteristic infrared absorption regions of the proposed reduced nitrogen species such as amides and amines are: NH stretching (3500–3050 cm^{-1}), $>$C$=$O absorption (1850–1650 cm^{-1}), NH deformation (1650–1500 cm^{-1}), and C—N stretching (1250–900 cm^{-1} for aliphatic and 1350–1300 cm^{-1} for aromatic). The interference could arise from the presence of the following species: NH_4^+ (v_3 and $v_2 + v_4$, 3500–3000 cm^{-1}), H_2O (v_2, near 1750–1550 cm^{-1}), CO_3^{2-} (v_3, near 1435 cm^{-1}), NH_4^+ (v_4, near 1400 cm^{-1}), NO_3^- (v_3, near 1360 cm^{-1}), and SiO_4^{2-}, PO_4^{2-}, and SO_4^{2-} (v_3, 1200–1000 cm^{-1}).

In order to determine the chemical structure of the species that were tentatively assigned to amines and amides, and in view of the difficulties when using infrared spectroscopy with ambient samples, it appeared that a more straightforward way would be to study samples produced under laboratory conditions, which are free of the interfering species. It is possible to generate the N_x species in surface reactions of ammonia with fine soot or graphite particles. We have employed two methods for their synthesis: one involves surface reactions of ammonia with combustion-generated soot particles in the presence of gaseous oxygen; the other method also involves an ammonia surface reaction but with finely ground graphite powder at low temperatures in the absence of oxygen. These "synthetic" N_x species have thermal behavior similar to that observed in ambient air particulates. To illustrate this, we shall first describe the results obtained with ambient samples collected in two different regions of California and then compare these results with those obtained with species synthesized by ammonia–particulate carbon reactions.

Figure 11 shows the results of one such measurement (17) for an ambient particulate sample collected in Pomona, California, during a moderate smog episode (October 24, 1972). The spectrum taken at a sample temperature of 25°C shows the presence of NO_3^-, NH_4^+, and N_x. At 80°C, the entire nitrate peak is lost, with a corresponding loss in the ammonium peak intensity. The shaded portion of the ammonium peak in

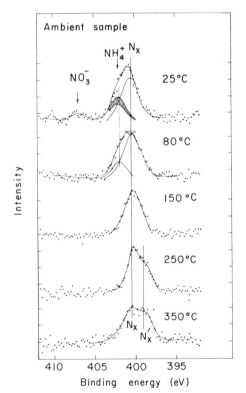

Figure 11. Nitrogen ($1s$) photoelectron spectrum of an ambient sample from Pomona, CA, as measured at 25, 80, 150, 250, and 350°C.

the 25°C spectrum represents the ammonium fraction volatilized between 25 and 80°C. The peak areas of the nitrate and the volatilized ammonium are approximately the same, indicating that the nitrate in this sample is mainly in the form of ammonium nitrate.The ammonium fraction still present at 80°C but absent at 150°C is associated with an ammonium compound more stable than ammonium nitrate, such as ammonium sulfate. At 150°C the only nitrogen species remaining in the sample is N_x. At 250°C the appearance of another peak, labeled N'_x, is seen. The intensity of this peak continues to increase at 350°C. The total $N_x + N'_x$ peak area at 150, 250, and 350°C remains constant, however, indicating that a part of the N_x is transformed into N'_x as a consequence of heating.

N'_x species will remain in the sample even if its temperature is lowered to 25°C, provided that the sample has remained in vacuum. However, if the sample is taken out of vacuum and exposed to the humidity of the air, N'_x will be transformed into N_x. It was concluded that N'_x species are produced by dehydration of N_x:

$$N_x \underset{+H_2O}{\overset{-H_2O}{\longleftarrow \longrightarrow}} N'_x \ .$$

Based on the described temperature behavior and on laboratory studies (17), N_x was assigned to a mixture of amines and amides. (N_x photoelectron peaks are broad indications of the presence of more than one single species.) Dehydration of the amide results in the formation of a nitrile N'_x.

Synthetic N_x species can be produced by surface reaction of ammonia with either graphite or combustion-produced soot particles (17). Results in Figure 12 show that N_x species produced by surface reactions of soot with NH_3 have the same kind of temperature dependence as the ambient sample mentioned. The spectrum taken at room temperature shows that most nitrogen species in this sample are of the N_x type. Heating the sample in vacuum to 150°C does not influence the line shape or intensity. At 250°C, however, the formation of N'_x is evident. Further transformation of N_x to N'_x occurs at 350°C.

Synthetic N'_x species will also remain unaltered even when the temperature is lowered to room temperature if the sample remains in vacuum.

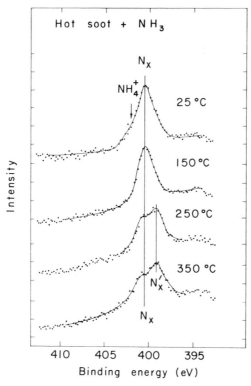

Figure 12. Nitrogen ($1s$) spectrum of a soot sample exposed to NH_3 at elevated temperature as measured at 25, 150, 250, and 350°C.

However, if this sample is taken out of vacuum and exposed to moisture, N_x' will be transformed back to the original N_x compound. The behavior of the synthetic N_x species is thus identical with the behavior of ambient species, such as those found in the Pomona sample.

Concerning the mechanism of this reaction, we speculated previously (17) that at elevated temperatures, ammonia will react by a nucleophilic substitution reaction with carboxyl groups associated with soot particle surfaces to produce an amide. This amide may become a nitrile by further dehydration. Alternative reaction with phenolic hydroxyl groups may yield an amine. Such reactions are to be expected because soot particles are composed not only of elemental carbon but also organic matter with 1–3% hydrogen and 5–15% oxygen by weight. Oxygen associated with soot particles is located in surface carbon–oxygen complexes, which typically are of carboxyl, phenolic hydroxyl, and quinone carbonyl type. Similar surface functional groups are easily formed when graphite is heated in an atmosphere of air and moisture.

Synthetic N_x species produced in surface reactions of soot with ammonia are not well suited for infrared spectroscopic determination of the amide structure because of interference in the $>C=O$ region. Therefore, it seemed advantageous to perform the infrared spectroscopic studies with samples that contain mostly particulate amines rather than a mixture of amines and amides. That such particulates exist in ambient air is suggested by the results shown in Figure 13, where the N(1s) ESCA spectra of a sample from Berkeley, California, (20) are shown.

Figure 13a shows the ESCA spectrum of this sample, recorded while the sample was exposed to vacuum and x-ray bombardment for about 1.5 h. The positions of the nitrate, ammonium, and amine peaks are indicated in the figure. The ammonium and nitrate peaks have a similar intensity, suggesting that these ions are present mostly as ammonium nitrate. The fraction of sulfate in the form of ammonium sulfate could also contribute to the ammonium peak. Figure 13b shows that a prolonged exposure of the sample to vacuum and x-ray bombardment has resulted in complete volatilization of ammonium and nitrate species. The only species that have remained in the sample are those that contain groups of the amino type. Heating of this sample to 350°C in vacuum resulted only in a slight change in the line shape (Figure 13c) in contrast to the ambient sample from Pomona, whose spectra at different sample temperatures are shown in Figure 11. This difference between the two samples could be tentatively explained if the Berkeley sample contains mostly amines, while the Pomona sample contains both amines and amides.

In the laboratory we have successfully produced (20) reduced nitrogen species that have ESCA spectra and temperature behavior similar to the behavior shown by the Berkeley sample mentioned earlier. The technique involves extensive grinding of graphite powder in ammonia in the absence of oxygen at room temperature. An examination by ESCA of the nitrogen species produced by such a surface reaction shows that these species are similar to those observed in the Berkeley sample, consisting mainly of species tentatively assigned as amines (Figure 14). Such synthetic samples were used for infrared studies.

The application of optical spectroscopy to study the structure of surface species on graphite is difficult because of its high absorption coefficient.

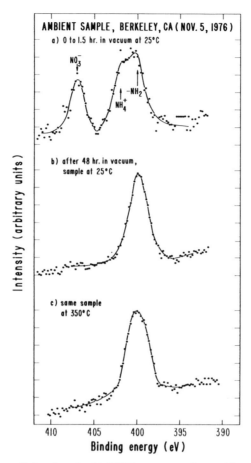

Figure 13. Nitrogen ($1s$) regions in the ESCA spectrum of an ambient sample collected in Berkeley, California. (*a*) Spectrum after the sample had been exposed to vacuum and x-ray bombardment for 1.5 h. (*b*) Spectrum of the same sample after it had been in vacuum for 48 h. (*c*) Spectrum of the same sample heated to 350°C.

The grinding technique employed enriches the concentration of surface nitrogen species. In order to help in assigning the vibrational frequencies, the samples were prepared by grinding graphite in both NH_3 and deuterated ammonia, ND_3. After grinding, the carbon particles were thoroughly mixed with KBr powder, pressed into pellets, and used for Fourier transform infrared analysis.

Figures 15*a* and 15*c* show infrared spectra (20) of the graphite particles after extensive grinding in an atmosphere of NH_3 and ND_3, with expansions of these spectra in Figures 15b and 15d. These infrared spectra suggest the occurrence of dissociative chemisorption of NH_3 on the carbon particle surface. Vibrational frequencies associated with the surface groups C—NH_2, C=N—H, C≡N, and C—H are observed in Figures 15a and 15b. The isotope shifts shown in Figures 15c and 15d support these assignments. Surface CNH_2 groups give rise to two bands near 3400 cm^{-1} that are attributed to symmetric and antisymmetric N—H stretching modes. These two bands should shift to 2500 cm^{-1} for CND_2. This shift is shown in Figures 15c and 15d. An NH_2 bending mode near 1580 cm^{-1} should shift to about 1200 cm^{-1} for the ND_2 group. However, a strong band due to the $k = 0$, E_{2g} phonon mode of the graphite lattice and/or a vibrational mode of the aromatic structure of graphite (21) also occurs at about 1580 cm^{-1}. Likewise, the C—N stretching mode vibrates at approximately 1200 cm^{-1} and appears in both the C—NH_2 and the C—ND_2 surface groups.

We have detected surface nitrogen groups indicating the dissociation of more than one bond in a molecule of ammonia. A band between 1600 and 1700 cm^{-1} could be assigned to immines (C=NH and C=N—C), a weak band at 2300 cm^{-1} to nitrile (C≡N), and one at 2180 cm^{-1} to isocyanide (—N^+≡C^-).

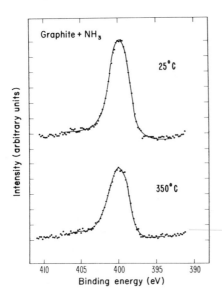

Figure 14.   Nitrogen (1s) spectrum of a sample produced by grinding graphite powder in an atmosphere of ammonia, measured at sample temperatures of 25 and 350°C.

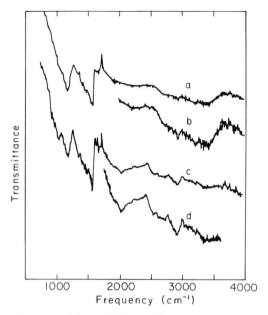

Figure 15. Infrared spectra of the graphite particles after extensive grinding in an atmosphere of NH_3 (*a*) and ND_3 (*c*). (*b*) and (*d*) are $2\times$ expansions along the ordinates of (*a*) and (*c*), respectively.

The evidence of the dissociative chemisorption of ammonia on carbon particle surfaces is also supported by the appearance of the C—D stretching band at 2050 cm^{-1}. The assignment of the C—H stretching is ambiguous because the C—H stretching is near 2900 cm^{-1} where a vibrational band appears on both NH_3 and ND_3 samples. This band could be the overtone and/or combination bands resulting from the strong absorption band between 1300 and 1600 cm^{-1}. There is also a band, possibly of the same nature, at 2700 cm^{-1} in both samples.

Acknowledgment

This work was supported by the Assistant Secretary for the Environment, Office of Health and Environmental Research, Pollutant Characterization and Safety Research Division of the U.S. Department of Energy under Contract No. DE-AC03-76SF00098.

REFERENCES

1. K. Siegbahn, C. Nordling, A. Fahlman, R. Nordberg, K. Hamrin, J. Hedman, G. Johansson, T. Bergmark, S. E. Karlsson, I. Lindgren, and B. J. Lindberg, *Nova Acta Regiae Soc. Sci. Ups.* **IV**, 20 (1967).

2. N. L. Craig, A. B. Harker, and T. Novakov, *Atmos. Environ.* **8**, 15 (1974).

3. T. Novakov, J. W. Otvos, A. E. Alcocer, and P. K. Mueller, *J. Colloid Interface Sci.* **39**, 225, 1972.

4. T. Novakov, S.-G. Chang, and R. L. Dod, in *Contemporary Topics in Analytical and Clinical Chemistry*, D. M. Hercules et al., Eds., Plenum, New York, 1977, p. 249.

5. T. Novakov, in *Analysis of Airborne Particles by Physical Methods*, H. Malissa, Ed., CRC Press, West Palm Beach, 1978, p. 191.

6. H. Malissa, H. Puxbaum, and E. Pell, *Fresenius Z. Anal. Chem.* **282**, 109 (1976).

7. H. Puxbaum, *Fresenius Z. Anal. Chem.* **299**, 33 (1979).

8. R. Dod et al., in *Atmospheric Aerosol Research FY-1979*, Lawrence Berkeley Laboratory report LBL-10735, 1980, p. 8–25; available from NTIS, Springfield, Virginia.

9. R. L. Dod, H. Rosen, and T. Novakov, in *Atmospheric Aerosol Research Annual Report 1977–78*, Lawrence Berkeley Laboratory report LBL-8696, p. 2; available from NTIS, Springfield, Virginia.

10. H. Rosen, A. D. A. Hansen, R. L. Dod, and T. Novakov, *Science* **208**, 741 (1980).

11. R. L. Dod and T. Novakov, paper presented at the American Chemical Society Symposium on Industrial Applications of Surface Analysis, New York, August 24–25, 1981; Lawrence Berkeley Laboratory report LBL-12592; to be published in the ACS Symposium Series.

12. B. R. Appel, J. J. Wesolowski, E. M. Hoffer, S. Twiss, S. Wall, S.-G. Chang, and T. Novakov, *Environ. Anal. Chem.* **4**, 169 (1976).

13. L. A. Gundel, S.-G. Chang, M. S. Clemenson, S. S. Markowitz, and T. Novakov, in *Nitrogenous Air Pollutants*, D. Grosjean, Ed., Ann Arbor Science, Ann Arbor, 1979, p. 211.

14. P. A. Russel, in *Proceedings, Carbonaceous Particles in the Atmosphere*, T. Novakov, Ed., Lawrence Berkeley Laboratory report LBL-9037, 1979, p. 133; available from NTIS, Springfield, Virginia.

15. S.-G. Chang and T. Novakov, in *Atmospheric Aerosol Research Annual Report 1976–77*, Lawrence Berkeley Laboratory report LBL-6819, 1977, p. 57; available from NTIS, Springfield, Virginia.

16. M. Clemenson, T. Novakov, and S. S. Markowitz, *Anal. Chem.* **51**, 572 (1979).

17. S.-G. Chang and T. Novakov, *Atmos. Environ.* **9**, 495 (1975).

18. P. T. Cunningham, S. A. Johnson, and R. T. Yang, *Environ. Sci. Technol.* **8,** 131 (1974).

19. P. T. Cunningham and S. A. Johnson, *Science* **191,** 77 (1976).

20. S.-G. Chang and T. Novakov, unpublished data; see also Ref. 15.

21. R. A. Friedel and L. J. E. Hofer, *J. Phys. Chem.* **74,** 2921 (1970).

CHAPTER

6

AN INTRODUCTION TO MULTIVARIATE ANALYSIS OF ENVIRONMENTAL DATA

PHILIP K. HOPKE

Institute for Environmental Studies
University of Illinois
Urbana, Illinois 61801

1. INTRODUCTION

Major efforts have been made in the recent past to substantially improve the analytical methodology applied to the study of environmental samples. The development of trace multielemental analysis using such techniques as instrumental neutron activation, proton- or photon-induced x-ray fluorescence, and inductively coupled plasma emission spectroscopy has permitted the determination of the concentrations of a large number of elements in a single sample. Similarly improved organic analytical procedures such as gas chromatography coupled with mass spectrometry and high-pressure liquid chromatography have greatly increased the number of organic compounds that can be identified and quantified even at trace levels. In combination, these procedures could determine a very large number of elemental and organic constituent concentrations in any given sample.

It has also become recognized that it is necessary to analyze a reasonably large number of samples in any monitoring effort in order to assess the wide variations in composition that an environmental system can possess. The combination of large numbers of samples that are analyzed for many constituent species yields data sets of such size and complexity that it is often difficult to understand the nature of the system being studied. Often the data are written into reports merely to languish on the shelves of the investigators.

This fate is also the common one for the vast quantities of data taken for the purpose of writing environmental impact statements. As Schindler (1) points out in his editorial in *Science*, the quality and utility of such efforts should be seriously questioned. At best they represent a collection of diverse data that is presented with little or no interpretation or description.

Many ecological studies have as their prime mission merely to collect "baseline" data to use as a reference against which future changes can be assessed. These studies could be made far more useful if some attempt were made to interrelate the measured parameters and to begin to develop a picture of how the system operates. It would be extremely valuable if the important relationships among the variables in the system could be ascertained and assist the experimenters in assessing what other kinds of data are required to develop additional insights for an ecosystem.

It is just now being realized by chemists that there are many possible uses for statistical analytical techniques in solving chemical problems. The physical sciences generally have been considered to deal only with systems in which all but one variable is held constant and the response of the system is determined as the remaining parameter is changed systematically. In many situations in a variety of current investigations, this basic method cannot be applied. For example, in the study of the chemistry of environmental systems, it is almost impossible to measure all the parameters that may affect the state of the system, and the control of any of these parameters is essentially impossible. There are many other situations in chemical problem solving in which the experimenter is confronted with a large body of data where there are undoubtedly multiple interactions or causalities giving rise to the observed values. Over the past few years there has been an increasing use of multivariate statistical techniques in solving chemical problems. Statistical analysis has been applied to gas chromatographic data (2, 3), nuclear magnetic resonance (4–6), mass spectra data (7–9), and infrared spectral data (10). Geological and biological studies have for a long time employed statistical methods for interpreting field data. However, these procedures have only recently been extensively applied in chemical studies of environmental systems (11–24). It is the purpose of this chapter to describe the basic principles of several of these techniques and to illustrate their application to environmental analytical chemistry.

2. STATISTICAL BACKGROUND

In determining the quantity of a particular element or compound in a specific sample at a definite time, the investigator has randomly removed a sample from a distribution of materials present in the environment. Then, by taking enough samples, the distribution of that particular variable in that kind of sample can be described by several parameters commonly used for that purpose including the mean value of the jth variable

$$\bar{x}_j = \frac{1}{n} \sum_{i=1}^{n} x_{ji} \tag{1}$$

and the second moment of the distribution or the *variance*

$$s_j^2 = \frac{1}{n-1} \sum_{i=1}^{n} (x_{ji} - \bar{x}_j)^2 \tag{2}$$

where n is the number of samples examined. The standard deviation is simply the square root of the sample variance.

For some statistical procedures, it is necessary to remove the effects of using different metrics in describing the various variables, and so the variables are put in standard form. First, the deviation is calculated by subtracting the mean value from each sample value.

$$d_{ji} = x_{ji} - \bar{x}_j \tag{3}$$

The standardized variable z_j can be calculated by dividing the deviation by the standard deviation

$$z_{ji} = \frac{d_{ji}}{s_j} = \frac{x_{ji} - \bar{x}_j}{s_j} \tag{4}$$

The standardized variable then has a mean value of zero and a standard deviation of unity; thus all standardized variables have the same metric.

The initial step in the analysis of the data generally requires the calculation of a function that can indicate the degree of interrelationship that exists within the data. Functions exist that can provide this measure between the two variables when calculated over all the samples or between the samples when calculated over all the variables. The most well known of these functions is the product-moment correlation coefficient. To be more precise, this function should be referred to as the correlation about the mean. The "correlation coefficient" between two variables x_j and x_k over all N samples is given by

$$r_{jk} = \frac{\sum_{i=1}^{n} (x_{ji} - \bar{x}_k)(x_{ki} - \bar{x}_k)}{[\sum_{i=1}^{n} (x_{ji} - \bar{x}_j)^2 \sum_{i=1}^{n} (x_{ki} - \bar{x}_k)^2]^{1/2}} \tag{5}$$

Utilizing the standardized variables, Eq. (5) simplifies to

$$r_{jk} = \frac{1}{N} \sum_{i=1}^{n} z_{ji} z_{ki} \tag{6}$$

There are several other measures that can also be utilized. These measures include covariance about the mean

$$c_{jk} = \sum_{i=1}^{n} d_{ji} d_{ki} \tag{7}$$

the covariance about the origin

$$c'_{jk} = \sum_{i=1}^{n} x_{ji} x_{ki} \tag{8}$$

and the correlation about the origin

$$r'_{jk} = \frac{\sum_{i=1}^{n} x_{ji} x_{ki}}{(\sum_{i=1}^{n} x_{ji}^2 \sum_{i=1}^{n} x_{ki}^2)^{1/2}} \tag{9}$$

The matrix of either the correlations or covariances, which is called the dispersion matrix, can be obtained from the original or transformed data matrices. The data matrices contain the data for the m variables measured over the n samples. The correlation about the mean is given by

$$(R_m) = (Z)(Z)^T \tag{10}$$

where $(Z)^T$ is the transpose of the standardized data matrix (Z). The correlation about the origin

$$(R_o) = (Z^o)(Z^o)^T = (XV)(XV)^T \tag{11}$$

where

$$z_{ji}^o = \frac{x_{ji}}{(\sum_{i=1}^{n} x_{ji}^2)^{1/2}}$$

is a normalized variable still referenced to the original variable origin and (V) is a diagonal matrix whose elements are defined by

$$v_{ij} = \delta_{ij}(\sum_{i=1}^{n} x_{ji}^2)^{-1/2} \tag{12}$$

The covariance about the mean is given as

$$(C_m) = (D)(D)^T \tag{13}$$

where $(D)^T$ is the transpose of the matrix of deviations from the mean calculated using Eq. (3). The covariance about the origin is

$$(C_o) = (X)(X)^T \tag{14}$$

the simple product of the data matrix by its transpose. As written, these product matrices would be of dimension $m \times m$ and would represent the pairwise interrelationships between variables. If the order of the multiplication is reversed, then the resulting $n \times n$ dispersion matrices contain the interrelationships between samples.

The relative merits of these functions to reflect the total information content contained in the data has been discussed in the literature (7, 25). Rozett and Petersen (7) argue that since many types of physical and chemical variables have a real zero, the information regarding the location of the true origin is lost by using the correlation and covariance about the mean, which include only differences from the variable mean. The normalization made in calculating the correlation from the covariances causes each variable to have an identical weight in the subsequent analysis. In mass spectrometry where the variables consist of the ion intensities at the various m/e values observed for the fragments of a molecule, the normalization represents a loss of information because the metric is the same for all the m/e values. In environmental studies where measured species concentrations range from the trace level (sub part per million) to major constituents at the percent level, the use of covariance may weight the major constituents too heavily in the subsequent analyses. The choice of dispersion function depends heavily on the nature of the parameters being measured.

Besides having a convenient measure of the interrelationship between two variables, it is also useful to develop procedures to describe the relationships between samples so that subsequently the samples can be grouped according to how similar or dissimilar they are to one another. One set of possible functions to describe the relationship between samples are the correlation and covariance function defined in the Eqs. (6)–(9) with the meaning of the indexes changed so that j and k refer to different samples and the summations are taken over the m variables in the system. The use of such functions will be explored later in this chapter.

An alternative approach is to define one of several geometrical measures. Consider an m-dimensional space. Each axis represents one of the variables in the system so that the values of all the measured variables for a single sample can be represented by a point. The distance between two points would be an indication of the dissimilarity between two samples since the larger the value of distance, the more dissimilar the two samples are from one another. One measure of distance is merely the extension of the simple Euclidean distance to an m-dimensional space. The Euclidean distance (ED) is given by

$$\text{ED}_{jk} = \left[\sum_{i=1}^{m} (x_{ji} - x_{ki})^2 \right]^{1/2} \tag{15}$$

Other functions have also been used including squared Euclidean distance (SED)

$$\text{SED}_{ik} = \sum_{i=1}^{m} (x_{ji} - x_{ki})^2 \tag{16}$$

mean character difference (MCD)

$$\text{MCD}_{jk} = \frac{1}{m} \sum_{i=1}^{m} (x_{ji} - x_{ki}) \tag{17}$$

mean Euclidean distance (MED)

$$\text{MED}_{ik} = \frac{1}{m} \left[\sum_{i=1}^{m} (x_{ii} - x_{ki})^2 \right]^{1/2} \tag{18}$$

and the mean squared Euclidean distance (MSED)

$$\text{MSED}_{jk} = \frac{1}{m} \sum_{i=1}^{m} (x_{ji} - x_{ki})^2 \tag{19}$$

An alternative approach is to consider a vector drawn from the origin to each of the n points. A measure of the similarity between two samples could be the cosine of the angle between their respective vectors (26). The cosine is given by

$$\cos \theta_{jk} = \frac{\sum_{i=1}^{m} (x_{ji})(x_{ki})}{(\sum_{i=1}^{m} x_{ji}^2 \sum_{i=1}^{m} x_{ki}^2)^{1/2}} \tag{20}$$

With only positive valued data, this measure has a range of 0–1. A value of 0.0 signifies there is nothing in common between the two samples, 1.0 shows identical samples, and 0.7071 (cos 45°) shows that the two vectors are about as similar as columns of random digits. The cos θ is clearly just the correlation about the origin between the jth and kth samples that can be given this geometrical interpretation (26). Now that the various interrelationship parameters have been defined, the mechanisms of the analytical methods can be discussed.

3. PRINCIPAL COMPONENT AND COMMON FACTOR ANALYSIS

The goal of factor and components analysis is to simplify the quantitative description of a system by determining the minimum number of new variables necessary to reproduce various attributes of the data. Factor analysis attempts to reproduce maximally the matrix of correlations, while principal components analysis attempts to reproduce the variance in the system. In the social science and educational literature, there is a substantial distinction made between factor and components analysis as discussed by Harmon (27). However, many of the differences are a matter of definition. These procedures reduce the original data matrix from one having m variables necessary to describe the n samples to a matrix with p components or factors ($p < m$) for each of the n samples. In permitting a parsimonious representation of data, it also permits the identification of the nature of the factors by giving the correlation coefficient between the original variables and the components. By examining which variables are highly correlated on a single component and with the investigator's understanding of the system under study, the nature of the factor can be deduced.

In both component and factor analysis, the value of the variable observed in the sample is assumed to be a linearly additive function of the contribution from each of the p causalities that actually govern the system's composition. For example, for airborne particulate matter, the amount of particulate lead in the air could be considered to be a sum of contributions from several sources including automobiles, incinerators and coal-fired power plants.

$$x_T(\text{Pb}) = x_1(\text{Pb}) + x_2(\text{Pb}) + x_3(\text{Pb}) + \cdots \qquad (21)$$

where x_T is the observed total particulate lead per unit volume, $x_1(\text{Pb})$ is the amount of lead per unit volume from source 1, $x_2(\text{Pb})$ is the amount per unit volume from source 2, and so forth for all of the p possible sources.

The individual source contributions can generally be thought of as a product of two cofactors: one that gives the amount of lead in the particles emitted by that specific source and the other that gives the mass of particulate matter per unit volume present that can be attributed to that source. Thus the $x_i(Pb)$'s can be rewritten as

$$x_i(Pb) = a_i(Pb)f_i \qquad (22)$$

Expanding this approach to all of the m variables that have been measured for the n samples gives

$$x_{ij} = a_{i1}f_{1j} + a_{i2}f_{2j} + \cdots + a_{ip}f_{2p} \qquad (23)$$

where x_{ij} is the value of the ith variable for the jth sample. Equation (23) is the model for principal components analysis. The major difference between factor analysis and components analysis as used in the social science literature is the requirement that common factors have significant values for a for more than one variable and an extra factor unique to each particular variable is added to the factors common to all of the variables. The factor model can be rewritten as

$$x_{ij} = a_{i1}f_{2j} + a_{i2}f_{2j} + \cdots + a_{ip}f_{pj} + d_iU_{ij} \qquad (24)$$

where U_{ij} is a factor unique to the ith variable calculated for the jth sample.

In the principle components analysis, components can be found with a strong relationship to only a single variable. This single variable component could also be considered to be the unique factor. Many of the traditional factor analysis texts (27) make much of the distinction. However, the two models really aim at the same result, and the differences are primarily differences in definition and semantics.

The model as expressed in Eq. (23) can be expressed in matrix notation as

$$(X) = (A)(F) \qquad (25)$$

where (X) is the $m \times n$ data matrix, (A) is the $m \times p$ factor loading matrix and (F) is the $p \times n$ factor score matrix. Factor analysis must permit determination of the number of terms required, p, and of one of the two matrices.

The resolution of the data matrix into the product of two cofactor matrices is accomplished by the diagonalization of one of the four dispersion matrices given in Eqs. (10), (11), (13), or (14) that represent the variable correlations. An analysis can also be made on the dispersion matrices that give the *sample* interrelationships. In many applications of factor analysis, the analysis was made on the matrix of correlations about the mean between the variables. In the terminology of Rozett and Peterson, these are R analyses. It is suggested by Rozett and Peterson (7) that it may be advantageous to analyze the matrix of correlations between samples. This is called a Q analysis. Both analyses aim for the same result: the determination of the (A) and (F) matrices. The R analysis calculated the (A) matrix first while the Q analysis determines the (F) matrix. The remaining matrix is calculated from the data using Eq. (25). In both cases the unmodified dispersion matrix is analyzed, and this procedure would be properly described as a principal components analysis. The diagonalization process for the R analysis is described in any book on standard factor analysis (27). The Q analysis is discussed by Malinowski and Howery (28). The diagonalization process compresses the information content of the data set into a limited number of eigenvalues

$$(Q)^{-1}(R)(Q) = (\Lambda) \qquad (26)$$

where (R) is the correlation matrix and Q and $(Q)^{-1}$, its inverse are found such that (Λ) is a diagonal matrix with the eigenvalues as the elements along the principal diagonal. The (Q) matrix contains the eigenvectors corresponding to each eigenvalue. In theory there should be a sharp distinction between the eigenvalues that contain the variance in the system due to the actual governing processes and that due to random error. In practice the choice of the proper number of factors to retain in the analysis is often quite difficult. It can be useful to plot the eigenvalues as a function of factor number and look for sharp breaks in the slope of the line (29). These breaks can provide an indication of the point of separation. Several commonly employed methods have been reviewed (25), and it appears that there are no uniformly applicable methods to determine the number of factors to be retained. Algorithms can be found to perform the eigenvalue analysis in the statistical program packages available on most computer systems. Because of the many approaches of factor analysis to analyzing data, care must be taken before using these library routines until the user has ascertained from the documentation that these routines do what he wants. In most statistical packages, there are built-in assumptions that the user may or may not desire. The most common assumption has to do with determining the number of factors to be used. Many routines consider only the eigenvectors having eigenvalues greater than 1.0 as significant factors unless the user explicitly sets other parameters, such as total included variance, to override the default on the eigenvalues. Although the eigenvalue-greater-than-one criterion sets a lower bound on the number of factors (30), it does not provide a simultaneous upper limit (31). In general, it appears that the best approach to principal component factor analysis is to set the parameters so that the computer codes will step through the analysis for various numbers of factors and then decide what number of factors should be considered. A useful aid in making this decision is discussed later.

To illustrate an R-mode principal components analysis process, a test data set will be analyzed. The data set was created assuming that a system exists that has four true causes, A, B, C, and D. For this system, 10 variables X1 to X10 are measured. Each of these variables is dependent on the four true causalities. The relationships between the sets of variables are given in Table 1. A data set of 100 samples was created by assigning random values to each of the four causal variables for each sample. Values

TABLE 1. Functional Dependence of the Ten Test Variables on the Four Causal Factors

	A	B	C	D
X1	0.25	0.25	0.25	0.25
X2	0.60	0.40	0.00	0.00
X3	0.00	0.10	0.90	0.00
X4	0.30	0.30	0.30	0.10
X5	0.10	0.30	0.40	0.20
X6	0.00	0.00	0.25	0.75
X7	0.50	0.00	0.50	0.00
X8	0.10	0.65	0.00	0.25
X9	0.10	0.10	0.70	0.10
X10	0.00	0.40	0.35	0.25

TABLE 2.   Correlation Matrix for the Test Data Set

	X1	X2	X3	X4	X5	X6	X7	X8	X9	X10
X1	1.000									
X2	0.827	1.000								
X3	0.750	0.425	1.000							
X4	0.976	0.886	0.773	1.000						
X5	0.965	0.709	0.863	0.950	1.000					
X6	0.737	0.292	0.561	0.579	0.725	1.000				
X7	0.848	0.774	0.810	0.893	0.807	0.480	1.000			
X8	0.835	0.775	0.442	0.820	0.816	0.529	0.477	1.000		
X9	0.849	0.539	0.984	0.855	0.922	0.653	0.873	0.534	1.000	
X10	0.929	0.652	0.795	0.899	0.981	0.733	0.677	0.877	0.850	1.000

of X1 through X10 were then calculated from these values. The correlation coefficients can be calculated between each of the pairs of variables or between pairs of samples. The coefficients of correlation about the mean for each pair of variables are calculated and are given in Table 2. The eigenvalue analysis is shown in Table 3. Clearly, in the case of the test data, the choice of four factors is very easily made since the only error in the data points is computational. Real data include errors from both sampling and analysis so that each value has uncertainty in it. The ei-

TABLE 3.   Eigenvalue Analysis of the Test Data Set

Number	Eigenvalue	Percent of Total Variance	Cumulative Percent of Variance
1	7.87	78.71	78.71
2	1.01	10.06	88.77
3	0.789	7.89	96.66
4	0.334	3.34	100.00
5	0.000	0.00	100.00
6	0.000	0.00	100.00
7	0.000	0.00	100.00
8	0.000	0.00	100.00
9	0.000	0.00	100.00
10	0.000	0.00	100.00

genvalues can be sorted into those representing true variation in the system resulting from the fundamental causalities in the system and those representing errors.

After the choice is made as to how many factors are to be retained, there still remains the problem of relating the resultant (A) matrix, called factor loadings, to terms meaningful to the system under study. The initially calculated factors are generally not interpretable in terms of physically real causalities and another transformation of these vectors is necessary in order to understand the nature of these factors. The factor loadings given in Table 4 bear little apparent relationship to the values in Table 1 and only through a rotation of the factor axes can these values be made to coincide. There is substantial controversy over the validity of axis rotation. Blackith and Reyment (32) strongly object to what they consider to be an arbitrary procedure and cite several references that agree with their position. Their purpose in utilizing multivariate analysis is to objectively classify objects into groups, and for that purpose, rotation does not help. In fact, a rotation that does not retain the orthogonality of the components would hinder such use. However, for the purpose of aiding the identification of the nature of the causal factors in a system, rotations have proven extremely useful (12–17). After the (A) matrix has been rotated the matrix of factor scores (F) can be calculated from the standardized data.

In applications of factor analysis to environmental chemical studies,

TABLE 4. Factor Loading Matrix Before Rotation for the Test Data Set

	A_1	A_2	A_3	A_4
X1	0.986	0.102	−0.054	0.119
X2	0.779	0.552	0.262	0.134
X3	0.843	−0.461	0.187	−0.202
X4	0.978	0.160	0.126	0.009
X5	0.990	−0.052	−0.083	−0.095
X6	0.705	−0.300	−0.533	0.356
X7	0.867	−0.085	0.440	0.214
X8	0.803	0.466	−0.316	−0.190
X9	0.916	−0.367	0.142	−0.076
X10	0.951	−0.000	−0.243	−0.190

the matrix of factors is generally rotated in such a way as to maximize the number of values that is close to either zero or unity. This rotation criterion is called "simple structure" (see the appendix in Reference 13) and a varimax rotation (33) may be used to achieve it. Simple structure may not be a very useful criterion for environmental applications since an element may be present in a sample because of several different causal forces. Therefore, factor loadings should not necessarily have values of either 0 or 1 but some intermediate value. The problem in inexact alignment resulting from the varimax rotation can be seen by examining the rotated factor loading matrix given in Table 5. The pattern that is now discerned looks very much like the coefficients used to generate the data (Table 1) except for the need to renormalize the sum of the coefficients to unity. The match of the component coefficients to the original ones is not perfect due to the error in the rotational model. It is clear that the components analysis has identified the proper number of causal factors and has given a good approximation of the numerical relationship between the variables and the components, but has not reproduced exactly the original relationships.

The varimax or other orthogonal rotation of this type attempts to repartition the variance in the system over all the factors retained. The examination of this variance *after* rotation can often be useful in determining the number of factors to retain. For this type of analysis, each of

TABLE 5. Orthogonally Rotated Factor Loading Matrix for the Test Data Set

	A_1 (C)	A_2 (B)	A_3 (A)	A_4 (D)
X1	0.473	0.526	0.558	0.431
X2	0.164	0.502	0.848	0.024
X3	0.936	0.210	0.188	0.207
X4	0.530	0.525	0.623	0.229
X5	0.638	0.569	0.364	0.367
X6	0.318	0.271	0.095	0.903
X7	0.654	0.093	0.724	0.195
X8	0.149	0.897	0.346	0.226
X9	0.862	0.253	0.309	0.309
X10	0.560	0.696	0.236	0.380

the original variables was standardized to have a variance of 1.0. Thus a rotated factor that has a variance less than 1.0 explains less variance than one of the original variables and therefore seems unlikely to be important enough to keep. Therefore, a criterion has been suggested (34) for determining the factors to be retained as those having a variance ≥ 1.0 after an orthogonal rotation. This approach has proven to be a useful rule of thumb for several data sets (34, 35).

One of the valuable uses of this type of analysis is in screening large data sets to identify errors (36). With the use of atomic and nuclear methods to analyze environmental samples for a multitude of elements, very large data sets have been generated. Because of the ease in obtaining these results with computerized systems, the elemental data acquired are not always checked as thoroughly as they should be, leading to some, if not many, bad data points. It is advantageous to have an efficient and effective method to identify problems with a data set before it is used for further studies. Principal component factor analysis can provide useful insight into several possible problems that may exist in a data set, including incorrect single values and some types of systematic errors.

There are four types of errors that may be observed in the factor analysis of a set of data: individual data point errors, bias errors, random variable errors, and errors due to the improper use of factor analysis. To illustrate how these data errors may be identified, a set of geological data from the literature (37) will be employed. This set of data was obtained by neutron activation analysis for 29 elements of 15 samples taken from Borax Lake, California, where a lava flow had once occurred. The samples obtained were formed by two different sources of lava that combined in varying amounts and thus formed a mixture of two distinct mineral phases. The data were acquired and analyzed to identify the elemental profiles of each source mineral phase.

The eigenvalue analysis for this data set is given in Table 6. In this case, there is a large break at two factors, and it is rather easy to identify the number of factors as the two that were anticipated. The varimax rotated factor loadings are given in Table 7. The geological data were then perturbed in various ways in order to demonstrate the error identification process.

Individual data point errors are the result of an error in a specific measurement for a specific sample. These errors can be the result of clerical mistakes or spurious determinations in the analytical process.

TABLE 6. Eigenvalues for the Unperturbed Geological Data Set

Number	Eigenvalue	Percent of Total Variance	Cumulative Percent of Variance
1	24.47	84.39	84.39
2	1.66	5.71	90.11
3	0.797	2.75	92.85
4	0.507	1.75	94.60
5	0.490	1.69	96.29
6	0.344	1.19	97.48
7	0.236	0.81	98.29
8	0.178	0.61	98.90
9	0.125	0.43	99.33
10	0.073	0.25	99.58
11	0.053	0.18	99.77
12	0.035	0.12	99.89
13	0.025	0.09	99.97
14	0.008	0.03	100.00
15	0.000	0.00	100.00
16	0.000	0.00	100.00
17	0.000	0.00	100.00
18	0.000	0.00	100.00
19	0.000	0.00	100.00
20	0.000	0.00	100.00
21	0.000	0.00	100.00
22	0.000	0.00	100.00
23	0.000	0.00	100.00
24	−0.000	−0.00	100.00
25	−0.000	−0.00	100.00
26	−0.000	−0.00	100.00
27	−0.000	−0.00	100.00
28	−0.000	−0.00	100.00
29	−0.000	−0.00	100.00

TABLE 7.   Factor Loadings for the
Unperturbed Geological Data Set

	Factor 1	Factor 2
Th	0.963	0.257
Sm	0.895	0.400
U	0.953	0.296
Na	0.786	0.439
Sc	−0.965	−0.250
Mn	−0.964	−0.257
Cs	0.957	0.280
La	0.905	0.332
Fe	−0.964	−0.255
Al	−0.928	−0.162
Dy	−0.017	0.859
Hf	0.727	0.382
Ba	−0.229	−0.805
Rb	0.948	0.129
Ce	0.955	0.259
Lu	0.775	0.294
Nd	0.886	0.196
Yb	0.912	0.077
Tb	0.919	0.054
Ta	0.848	0.371
Eu	−0.966	−0.242
K	0.890	0.262
Sb	0.600	0.676
Zn	−0.923	−0.289
Cr	−0.970	−0.220
Ti	−0.965	−0.237
Co	−0.966	−0.247
Ca	−0.966	−0.156
V	−0.945	−0.240

Since an error is limited to a few samples in the data set, most of the variance associated with that variable is resident in one specific sample. A decimal point error is an example of an individual data point error. To simulate a decimal point error, the geological data described above was altered so that one of the 15 thorium values was off by a factor of 10. The data were processed by principal component analysis to produce factor loadings and factor scores for 2–10 factors. The factor loadings were then examined starting with the two-factor solution to determine the nature of the identified factors. For the two-factor solution, the first factor has many variables highly correlated to it, while the second factor is highly correlated to Dy, Ba, and Sb. For the three-factor solution, the preceding factors were again observed in addition to a factor containing most of the variance in thorium and little of any other variable as shown in Table 8. This factor that reflects a high variance for only a single variable shows one of the typical characteristics of a possible data error. To further investigate the nature of this factor, the factor scores of the three-factor solution were calculated and are given in Table 9. These scores are in a standardized form. An average value is 0 while a value of 1.0 is one standard deviation above average and −1.0 is one standard deviation below the mean value. There is a very large value for the score of factor 3 for sample 1, indicating that most of the variance was due to sample 1, which is the sample with the altered thorium value. Having made these observations, the raw data must be reexamined to decide if there is an explanation for the observed abnormality. It must be recognized that these points may represent a single rare event that is physically real. For example, Alpert and Hopke (19) identified three of 100 airborne particulate samples that were strongly affected by Fourth of July fireworks. Dattner and Jenks (38) identified such a factor that is found in Texas airborne particle composition data representing those few days during the winter when residual oil is burned to supplement natural gas combustion. Thus real isolated events are possible. However, in the case where there is no obvious explanation, the points may be recognized as erroneous and appropriate corrective action taken. Even when the points are thought to be real, it is generally useful to eliminate them from the data set and repeat the analysis.

If there were no physical explanation for the unique thorium contribution in the preceding example and the factor scores were small for all of the samples, then the error may have been distributed throughout the data for that variable. An example of this would be the presence of a variable dominated by random noise in the data set. To demonstrate how a random variable error may be identified, an additional variable X1 was added to the geological data set. The values for X1 were generated using a normally distributed random number generator to produce values with a mean of 5. and standard deviation 0.5. Using the procedure described

**TABLE 8.   Factor Loadings of the Geological Data with a
Decimal Point Error**

Element	Factor 1	Factor 2	Factor 3
Th	0.030	0.046	0.972
Sm	0.899	0.360	0.155
U	0.959	0.264	0.090
Na	0.778	0.362	0.342
Sc	−0.971	−0.221	−0.063
Mn	−0.970	−0.226	−0.075
Cs	0.962	0.247	0.093
La	0.904	0.286	0.181
Fe	−0.970	−0.226	−0.064
Al	−0.937	−0.155	0.034
Dy	−0.002	0.828	0.208
Hf	0.708	0.288	0.481
Ba	−0.262	−0.834	0.095
Rb	0.939	0.074	0.204
Ce	0.956	0.220	0.128
Lu	0.771	0.266	0.173
Nd	0.878	0.142	0.211
Yb	0.925	0.084	−0.119
Tb	0.923	0.024	0.032
Ta	0.837	0.303	0.334
Eu	−0.972	−0.212	−0.068
K	0.908	0.263	−0.092
Sb	0.621	0.674	0.024
Zn	−0.927	−0.247	−0.123
Cr	−0.976	−0.191	−0.058
Ti	−0.972	−0.211	−0.039
Co	−0.971	−0.216	−0.073
Ca	−0.963	−0.119	−0.121
V	−0.948	−0.208	−0.091

earlier, factor analysis identified the random variable as can be seen in
Table 10. Again, the same basic characteristics were observed in the
factor, that is, a large loading for variable X1 and little contribution for
the other variables. If one were to look at the factor scores for this case,
the distribution of the values for factor 3 would not identify a specific

**TABLE 9. Factor Scores of Geological Data with a
Decimal Point Error**

Sample	Factor 1	Factor 2	Factor 3
1	-0.179	0.101	3.51
2	-0.240	2.63	-0.591
3	0.037	1.06	-0.027
4	0.846	-0.517	-0.462
5	0.753	-0.560	0.011
6	0.782	-0.520	-0.029
7	0.900	-0.333	-0.426
8	0.645	0.522	-0.387
9	1.23	-1.53	-0.162
10	0.804	0.662	-0.074
11	-2.08	-1.02	-0.601
12	-1.58	-0.236	0.148
13	-1.12	-0.376	-0.500
14	-0.679	-0.486	-0.290
15	-0.109	0.613	-0.120

sample. Thus the problem with variable X1 is distributed over all the samples. If there is no physical reason for this variable being unique, then the investigator should explore his or her analytical method for possible errors.

Similar to random variable errors, bias errors may be found throughout a data set. Bias errors can be caused by:

1. Interference between two analytic species on the analytical instrument that measures their concentration.
2. Contamination or loss in the chemical manipulation of the sample such as extraction or precipitation.
3. Chemical changes during the time delay from sample collection to sample analysis.

To illustrate a bias error, two correlated random variables Y1 and Y2 were generated and added to the geological data set. The Y1 values were produced in the same manner as the X1 values described earlier. The Y2 values were generated by dividing the Y1 value by 3 and adding a normally distributed random error with a standard deviation of 0.15 to the result. After factor analysis, the factor loadings suggest the possibility of a problem in the data, as can be seen in Table 11. The large correlation between variables Y1 and Y2 for factor 3 along with the absence of any other contributing variable was very obvious.

TABLE 10.   Factor Loadings of the Geological Data with a Random Variable X1

Element	Factor 1	Factor 2	Factor 3
Th	0.966	0.245	−0.045
Sm	0.903	0.389	−0.002
U	0.955	0.286	−0.058
Na	0.832	0.340	0.153
Sc	−0.968	−0.236	0.048
Mn	−0.966	−0.243	0.060
Cs	0.958	0.270	−0.063
La	0.915	0.311	0.005
Fe	−0.966	−0.242	0.056
Al	−0.888	−0.223	0.296
Dy	0.048	0.757	0.341
Hf	0.728	0.362	−0.127
Ba	−0.231	−0.841	−0.050
Rb	0.963	0.078	−0.011
Ce	0.948	0.260	−0.122
Lu	0.710	0.408	−0.439
Nd	0.909	0.137	0.041
Yb	0.865	0.146	−0.343
Tb	0.944	−0.010	0.058
Ta	0.847	0.370	−0.059
Eu	−0.968	−0.226	0.074
K	0.880	0.277	−0.097
Sb	0.596	0.690	−0.100
Zn	−0.943	−0.241	−0.032
Cr	−0.975	−0.200	0.042
Ti	−0.968	−0.223	0.049
Co	−0.969	−0.233	0.050
Ca	−0.958	−0.154	0.116
V	−0.941	−0.232	0.109
X1	−0.006	0.290	0.856

TABLE 11.   Factor Loadings of the Geological Data with
Two Correlated Random Variables Y1 and Y2

Element	Factor 1	Factor 2	Factor 3
Y1	0.133	−0.160	0.888
Y2	−0.116	−0.010	0.857
Th	0.976	0.202	−0.004
Sm	0.916	0.352	−0.003
U	0.968	0.242	−0.009
Na	0.813	0.382	−0.145
Sc	−0.978	−0.193	0.027
Mn	−0.978	−0.200	0.018
Cs	0.972	0.224	−0.011
La	0.920	0.288	0.067
Fe	−0.977	−0.198	0.027
Al	−0.933	−0.117	−0.143
Dy	0.031	0.858	−0.066
Hf	0.742	0.361	0.229
Ba	−0.277	−0.783	0.044
Rb	0.956	0.066	−0.095
Ce	0.967	0.208	0.057
Lu	0.785	0.270	0.303
Nd	0.898	0.136	−0.048
Yb	0.917	0.020	0.050
Tb	0.924	−0.009	−0.136
Ta	0.862	0.340	0.189
Eu	−0.980	−0.182	0.049
K	0.904	0.210	−0.004
Sb	0.646	0.616	−0.273
Zn	−0.938	−0.234	−0.022
Cr	−0.982	−0.163	0.025
Ti	−0.977	−0.180	0.013
Co	−0.979	−0.191	0.017
Ca	−0.972	−0.103	0.041
V	−0.956	−0.188	−0.033

The question now arises, What if the two interfering variables are also correlated with other variables in the data? To investigate this, the concentrations of two uncorrelated elements, thorium and europium, were altered to simulate an interference. Europium values were altered by adding 10% of the thorium value plus a normally distributed random error about thorium with a 5% standard deviation. The thorium values were altered similarly except that 10% of the europium value was subtracted rather than added. As before, the problem in the data set was identified (see Table 12) in factor 3 by characteristics similar to those previously described. Again, this unusual factor establishes the possibility of a problem, and it is up to the researcher to identify the nature of the difficulty.

TABLE 12. Factor Loadings of the Geological Data with Errors Present in Two Previously Uncorrelated Variables, Th and Eu

Element	Factor 1	Factor 2	Factor 3
Th	0.965	0.234	0.091
Sm	0.889	0.383	0.093
U	0.956	0.275	0.087
Na	0.821	0.391	−0.109
Sc	−0.969	−0.226	−0.067
Mn	−0.967	−0.233	−0.077
Cs	0.958	0.259	0.093
La	0.904	0.309	0.129
Fe	−0.968	−0.232	−0.069
Al	−0.897	−0.165	−0.290
Dy	0.032	0.827	−0.198
Hf	0.712	0.370	0.253
Ba	−0.231	−0.820	−0.102
Rb	0.964	0.091	−0.021
Ce	0.947	0.242	0.160
Lu	0.717	0.324	0.498
Nd	0.899	0.156	0.024
Yb	0.882	0.081	0.264
Tb	0.944	0.009	−0.095
Ta	0.831	0.360	0.256
Eu	0.035	−0.178	0.935
K	0.887	0.248	0.099
Sb	0.626	0.669	−0.082
Zn	−0.932	−0.253	−0.070
Cr	−0.976	−0.194	−0.059
Ti	−0.968	−0.213	−0.076
Co	−0.969	−0.223	−0.075
Ca	−0.968	−0.137	−0.056
V	−0.942	−0.219	−0.118

Now, consider the situation where the two interfering variables are correlated as well as interfering with each other. Again the data set was altered as described earlier, except that the samarium values were substituted for the europium values. The resulting factor loadings are given in Table 13. The problem appears in factor 2, but it is not obvious that

TABLE 13. Factor Loadings of the Geological Data with Errors Present in Two Previously Correlated Variables, Th and Sm

Element	Factor 1	Factor 2	Factor 3
Th	0.854	0.472	0.206
Sm	0.414	0.810	−0.028
U	0.860	0.428	0.268
Na	0.831	0.132	0.412
Sc	−0.883	−0.405	−0.224
Mn	−0.879	−0.412	−0.230
Cs	0.862	0.431	0.252
La	0.811	0.427	0.297
Fe	−0.880	−0.408	−0.229
Al	−0.732	−0.616	−0.134
Dy	0.106	−0.178	0.857
Hf	0.624	0.421	0.340
Ba	−0.108	−0.340	−0.796
Rb	0.927	0.273	0.104
Ce	0.835	0.483	0.226
Lu	0.486	0.779	0.257
Nd	0.861	0.279	0.165
Yb	0.720	0.597	0.054
Tb	0.936	0.187	0.034
Ta	0.706	0.516	0.330
Eu	−0.890	−0.391	−0.217
K	0.783	0.435	0.242
Sb	0.542	0.282	0.678
Zn	−0.877	−0.339	−0.257
Cr	−0.896	−0.389	−0.194
Ti	−0.882	−0.408	−0.211
Co	−0.882	−0.411	−0.220
Ca	−0.875	−0.409	−0.132
V	−0.846	−0.436	−0.210

it is a difficulty since there are high values for some other elements. For these data, the researcher might be able to identify the problem concerning samarium if he or she has sufficient insight into the true nature of the data. In this case, it is known there are only two mineral phases. The existence of three factors indicates that either an error has been made in assuming only two phases or there are errors in the data set. Thus knowledge of the nature of the system under study is needed to find this type of error. The two variables involved could also be an indication of a problem. If two variables that would not normally be thought to be interrelated appear together in a factor, it could indicate a correlated error.

In both of the preceding examples, there were two interfering variables present, either thorium and samarium or thorium and europium. Potential problems in the data were recognized after using factor analysis, that is, samarium and europium. However, no associated problem was noticed with thorium in either case because of the relative concentrations of the two variables. Since thorium was much less sensitive to a change (because of its large magnitude relative to samarium or europium), the added errors in thorium were interpreted as unnoticeable increases in the variance of the thorium values. For an actual common interference, consider the Mn–Mg interference problem in neutron activation analysis. The gamma rays used to determine these elements have energies of 846.7 and 844.0 keV, respectively, and frequently their peaks overlap. If the spectral analysis program used had a problem properly separating the Mn and Mg peaks, the factor analysis would probably identify the problem as being in the Mg value since the sensitivity of Mg to neutron activation analysis is so much less than that of Mn. However, if the actual levels of Mn were so low that the Mg peak was of the same size as the Mn peak, then the problem could show up in both variables.

The fourth error type is due to the improper use of the factor analysis technique. If too many factors are retained in the analysis, the extra factors add noise to the system and cause the analysis to poorly represent the correlations in the data. These error factors are characterized by their lack of identifiable physical significance and may be eliminated by the proper choice of the number of factors as described earlier.

Thus an important first step in interpreting a data set is to perform an R-mode principal components analysis to assess the quality of the data and to identify any obvious problems with it. These problems should be eliminated before proceeding further.

This type of factor analysis with standardized variables and results has certain advantages in that it is able to combine different types of variables in the analysis, to identify variance in the data that arises from sampling and/or analytical procedure errors, and to provide a perspective on the data without any *a priori* knowledge of the system under study. This approach has the disadvantage of providing results in a standardized form. The principal components model as used earlier has the form

$$z_{ji} = \frac{x_{ji} - \bar{x}_j}{\sigma_j} = \sum_{k=1}^{p} a_{jk} f_{ki} \tag{27}$$

rather than that of Eq. (23), so that if the original data x_{ji} are to be calculated, the equation becomes

$$x_{ji} = \bar{x}_j + \sum_{k=1}^{p} \sigma_j a_{jk} f_{ki} \tag{28}$$

This equation partitions the differences from the mean value rather than partitioning the value of the variable itself. In many systems the value of a given variable should be a linearly additive sum of independent contributions, and a procedure to obtain a model like that in Eq. 23 is desirable. This approach is possible using the matrix of correlations about the origin as the dispersion matrix. In this case the standardization only involves division of the variable by the square root of the sum of the variable squared so that the resulting model can be written as

$$z_{ji}^o = \frac{x_{ji}}{(\sum_{i=1}^{n} x_{ji}^2)^{1/2}} = \sum_{k=1}^{p} a_{jk} f_{ki} \tag{29}$$

so that the original values can be directly partitioned

$$x_{ji} = \left(\sum_{i=1}^{n} x_{ji}^2 \right)^{1/2} \sum_{k=1}^{p} a_{jk} f_{kj}$$

$$= \sum_{k=1}^{p} a_{ji}' f_{kj} \tag{30}$$

where $a_{jk}' = a_{jk}(\sum_{j=1}^{n} x_{ji}^2)^{1/2}$.

This approach has been widely used by Malinowski and his co-workers (4–6, 9, 28) for a variety of physical chemical studies.

Another extensive use of this form of factor analysis has been by Hopke and co-workers in their source identification and quantitative mass apportionment of airborne particulate (18, 19, 39), of settled street dust (40) and of geological samples including coal (41–44). To illustrate this approach the geological data set described earlier will be analyzed using a Q-mode factor analysis procedure where the starting point is the matrix of correlations between samples. In these data, it is already known that two mineral phases are present, and this result is reflected in the various tests of the number of factors after the diagonalization of the matrix of correlations about the origin that is presented in Table 14.

There are several indicators of two factors being present. The data is reproduced using the first factor and then compared to the original data point by point using root-mean-square (RMS) error, chi-square, the Exner function (45), and average percent error. The data is then reproduced using the first two factors and so on. There is clearly a substantial improvement in the fit by using two factors, but there is not a further significant improvement by including additional factors. Two factors are strongly indicated.

Another important difference of this form of factor analysis is in the rotation step. Since the eigenvectors can be transformed to reflect the unnormalized variables in the system, it is possible to rotate a factor axis toward a target vector that represents a possible controlling agent of the system under study.

$$X = ARR^{-1}F \tag{31}$$
$$= A'F'$$

where A' is the matrix of rotated vectors that might represent the elemental compositions of the two phases present in the lava dike under study. The columns of the R matrix are obtained from a least-squares fit of a factor axis to a suggested phase composition or test vector b by

$$r = (A^T W A)^{-1} A^T W b \tag{32}$$

where A^T is the transpose of the factor loading matrix and W is a weighting matrix. In this case an elemental concentration profile of a dacite or an

TABLE 14. Reproduction Summary for the Geological Data Set

Factor	Eigenvalue	RMS	Chi-square	Exner	Error
1	1.4427E + 01	170.00	863.	0.2101	39.84
2	5.5359E − 01	28.87	33.	0.0384	7.57
3	1.2793E − 02	17.02	30.	0.0222	7.13
4	6.0663E − 03	3.60	21.	0.0049	4.55
5	2.9430E − 04	0.81	17.	0.0011	4.25
6	1.4198E − 05	0.21	20.	0.0003	3.85
7	6.3488E − 07	0.13	17.	0.0002	3.30
8	2.2905E − 07	0.09	20.	0.0001	2.33

obsidian could be tested to see if they are good representations of the factors in the system. Roscoe and Hopke (41) have shown that the rotation process can be improved through the use of the weight matrix and have suggested several different methods for weighting the fit.

Another important finding (41) is that the profiles of the mineral phases could be deduced by an iterative process from a very simple set of initial test vectors. These simple vectors consist of a value of 1.0 for one variable and 0.0 for all the others. The set of these unique vectors can be put into Eq. (32) and an r determined. From $r'b$, a new test vector b' can be calculated and used as the input vector to the next step. This process can be repeated until the average differences between elements in b and b' are sufficiently small ($<10^{-4}$). It was found that for the geological data set, excellent agreement, as shown in Table 15, could be obtained between the resultant profiles and the compositions of the two phases originally reported (37). This result indicates that a great deal of understanding of the system can be obtained from very limited *a priori* information.

After the factor analysis solution to a data set is achieved, it would be useful if the error in the analysis could be calculated. Two methods for the calculation of the uncertainties in the (A) and (F) matrices are available: the "jackknife" method (46, 47), which can be cumbersome and time-consuming, and a rapid method that has been previously described by Clifford (48). These methods have been compared by Roscoe and Hopke (42).

The jackknife method uses multiple determinations of A' obtained by repeating the complete factor analysis several times with a reduced data set where one observation has been deleted each time. Therefore, to ex-

TABLE 15. Concentrations of Elements in the Two Phases Present in the Geological Data Set

	Phase A		Phase B	
Element	Factor Analysis	Original Report	Factor Analysis	Original Report
Th	17.9	17.6	1.84	2.5
Sm	5.85	5.85	4.61	4.6
U	6.74	6.7	.55	.72
Na	27700.	27600.	24600.	24300.
Sc	4.47	4.76	38.9	37.5
Mn	122.	134.	1420.	1370.
Cs	16.1	15.9	0.	.50
La	23.0	22.7	12.5	12.5
Fe	6060.	6700.	80000.	77000.
Al	65000.	65000.	84300.	84000.
Dy	7.72	7.80	7.39	7.1
Hf	3.80	3.8	3.18	3.2
Ba	87.2	—[a]	135.	—[a]
Rb	250.	245.	5.66	27
Ce	53.4	52.8	27.2	27.6
Lu	0.52	0.52	0.42	0.43
Nd	26.9	25.6	15.3	18.
Yb	4.08	4.0	2.69	2.7
Tb	1.06	1.05	0.80	0.81
Ta	1.18	1.18	0.48	0.52
Eu	0.09	0.10	1.63	1.6
K	46400.	46000.	2550.	8000.
Sb	1.13	1.1	0.20	0.3
Zn	42.8	45.	118.	114.
Cr	7.96	9.4	186.	178.
Ti	235.	300.	9360.	9000.6
Co	0.	0.40	41.6	39.6
Ca	9500.	10400.	78800.	75000.
V	6.96	9.	301.	300.

[a] Not reported by Bowman et al. (38)

amine all of the possible combinations, the factor analysis must be done one time more than the number of observations in the study. The source profiles of the modified data, A_i', can be determined and used to calculate a better source profile in addition to an error in that profile. This determination is made by first forming a "pseudovalue" (17):

$$A_i^* = NA' - (N - 1)A_i' \tag{33}$$

where N is the number of observations in the total data set. The mean of the pseudovalues will give a better estimate of the source profiles.

$$\overline{A}_i^* = \frac{1}{N} \sum A_i^* \tag{34}$$

The variance s^2 of A^* can then be calculated

$$s^2 = \frac{\sum (A_i^*)^2 - (\sum A_i^*)^2/N}{[N(N - 1)]} \tag{35}$$

A calculational method is described by Clifford (48) and relies on the error obtained from comparing the reproduced data to the raw data. The matrix of differences between the original and calculated data are formed a row or column at a time

$$E = X(obs) - X(calc) \tag{36}$$

An average variance of the data in that row or column is then calculated by

$$\sigma^2 = \frac{E'WE}{m - n} \tag{37}$$

where W is the weight matrix previously used in the analysis, m is the number of variables, and n is the number of factors used. To apportion the total variance in the data, the matrix

$$P = (A'WA)^{-1} \tag{38}$$

is formed. The errors in the calculated results may be estimated by

$$\sigma(Y_i) = \sigma(P_{ii})^{1/2} \tag{39}$$

where $\sigma(Y_i)$ is the estimated error in parameter Y_i. Further details may be found in the literature (42). The results obtained for the elemental profiles of the mineral phases with the jackknife and calculational methods of error analysis are compared to those obtained from a simple linear regression analysis in Table 16. The results obtained by both the jackknife and calculational techniques are in good agreement with each other and with the linear regression results. Since the calculational method is both quicker and easier to use than the jackknife method, it has an inherent advantage.

A characteristic of both the jackknife and the calculational error techniques is that the error for the entire analysis is assumed to be resident in the parameters for which one is calculating the errors. For this reason, both techniques yield overestimates of the actual error.

These results illustrate the utility of several factor analysis models in both screening large data sets and quantitatively resolving them into appropriate components with estimates of the uncertainties in the results.

4. CLUSTER ANALYSIS

In many areas of study there is a need to be able to objectively group objects that have characteristics similar to one another. Cluster analysis is a technique that makes such groupings and displays the pattern of groups so that the investigator may see both the relationship between objects as well as between the groups of objects. The first large-scale efforts in this area came in sorting species into consistent sets of taxa, and a large literature has been developed in numerical taxonomy (49). Another area of increasing use of cluster analysis is in the sorting of archeological artifacts into groups having similar origins in terms of starting materials and/or their methods of fabrication (50–55). The application of cluster analysis to environmental problems will be discussed after the introduction to the principles of the technique.

The basic concepts of clustering can be best considered in geometric terms. In the space whose dimensionality is defined by the number of

TABLE 16. Comparison of Source Mineral Profiles and Their Associated Errors for the Borax Lake Geological Data

	Mineral Phase A			Mineral Phase B		
	Calculational Method	Jackknife Method	Linear[a] Regression	Calculational Method	Jackknife Method	Regression
Th	17.9 ± 0.3	17.9 ± 0.1	17.6 ± 0.1	1.8 ± 0.9	1.4 ± 0.6	2.5 ± 0.3
Sm	5.85 ± 0.06	5.86 ± 0.04	5.85 ± 0.03	4.61 ± 0.18	4.56 ± 0.11	4.60 ± 0.11
U	6.74 ± 0.09	6.75 ± 0.05	6.70 ± 0.04	0.55 ± 0.28	0.37 ± 0.23	0.72 ± 0.14
Na	27,700 ± 300.	27,700 ± 200.	27,600. ± 200.	24,600. ± 1,000.	24,400. ± 900.	24,300. ± 600.
Sc	4.48 ± 0.26	4.48 ± 0.04	4.76 ± 0.06	38.9 ± 0.9	39.8 ± 0.9	37.5 ± 0.2
Mn	122. ± 10.	122. ± 3.	134. ± 3.	1,410. ± 40.	1,450. ± 60.	1,369. ± 10.
Cs	16.1 ± 0.2	16.1 ± 0.2	15.9 ± 0.1	0.00 ± 0.66	0.42 ± 0.43	0.5 ± 0.4
La	23.0 ± 0.4	23.0 ± 0.2	22.7 ± 0.3	12.5 ± 1.3	11.9 ± 1.2	12.5 ± 1.0
Fe	6,060. ± 550.	6,050. ± 100.	6,700.	80,000. ± 1,800.	81,900. ± 2,100.	77,000.
Al	65,000. ± 600.	65,100. ± 600.	65,000. ± 640.	84,300. 1,800	84,800. ± 1,800.	84,000. ± 2,100.
Dy	7.72 ± 0.14	7.73 ± 0.17	7.80 ± 0.20	7.39 ± 0.47	7.30 ± 0.39	7.10 ± 0.65
Hf	3.80 ± 0.05	3.81 ± 0.04	3.80 ± 0.05	3.19 ± 0.16	3.15 ± 0.17	3.20 ± 0.16
Ba	87.2 ± 8.9	87.2 ± 10.0	86. ± 11[b]	130. ± 30.	140. ± 30.	140. ± 37.[b]
Rb	250. ± 7.	250. ± 7.	245. ± 8.	6. ± 22.	0. ± 12	27. ± 26.

Ce	53.4 ± 0.8	53.5 ± 0.4	52.8 ± 0.5	27.2 ± 2.4	26.2 ± 2.2	27.6 ± 1.6
Lu	0.522 ± 0.008	0.523 ± 0.006	0.52 ± 0.007	0.419 ± 0.024	0.416 ± 0.023	0.43 ± 0.022
Nd	26.9 ± 0.6	26.9 ± 0.4	25.6 ± 0.5	15.3 ± 1.8	15.1 ± 1.6	18.0 ± 1.6
Yb	4.08 ± 0.07	4.09 ± 0.06	4.00 ± 0.06	2.69 ± 0.23	2.65 ± 0.17	2.07 ± 0.20
Tb	1.06 ± 0.01	1.06 ± 0.01	1.05 ± 0.01	0.795 ± 0.031	0.791 ± 0.024	0.810 ± 0.036
Ta	1.18 ± 0.03	1.18 ± 0.03	1.18 ± 0.03	0.49 ± 0.10	0.47 ± 0.08	0.52 ± 0.10
Eu	0.0905 ± 0.0116	0.0900 ± 0.0058	0.100 ± 0.007	1.63 ± 0.04	1.67 ± 0.06	1.60 ± 0.03
K	46,400. ± 800.	46,400. ± 1100.	46,000 ± 1,400.	2,600. ± 2,500.	2,100. ± 4,500.	8,000. ± 4,800.
Sb	1.13 ± 0.06	1.13 ± 0.06	1.10 ± 0.08	0.20 ± 0.20	0.16 ± 0.11	0.30 ± 0.25
Zn	42.8 ± 2.0	42.8 ± 1.7	45.0 ± 1.9	118. ± 7.	120. ± 9.	114. ± 6.
Cr	8.0 ± 1.7	7.93 ± 0.78	9.40 ± 0.94	186. ± 6.	190. ± 6.	178. ± 3.
Ti	240. ± 100.	236. ± 30.	300. ± 55.	9,360. ± 330.	9,570. ± 300.	9,000. ± 180.
Co	0.00 ± 0.31	0.0010 ± .00358	0.40 ± 0.10	41.6 ± 1.1	42.6 ± 1.1	39.6 ± 0.3
Ca	9,510. ± 770.	9,490. ± 810.	10,400. ± 1,500.	78,700. ± 2,600.	80,400. ± 4,200.	75,000 ± 4,900.
V	7.0 ± 5.0	6.4 ± 4.2	9. ± 6.	301. ± 17.	313. ± 31.	300. ± 21.

[a] Errors in the linear regression coefficients were not reported in the original work (20). Errors reported here were produced from the data by the authors. It should be noted that all parameters were found by a regression comparing elemental concentrations to those assumed for iron in the two phases.

[b] Values for barium were not reported in the original work, but were constructed here by the authors.

251

variables, each sample is represented by a point. It is then of interest to see which points are physically close enough to one another to form a group or cluster, just as the stars in the night sky are grouped into constellations. However, we would like a quantitative measure of cluster membership as well as a definition of the relationships between groups. Several quantitative measures of distances between points or measures of sample dissimilarity have been defined previously.

It is essential that the variables be standardized before calculating the dissimilarity measures. If the original variables are used, then trace species will span a much smaller size space than minor or major constituents. Thus the major species will dominate the distance calculation. It is the variation within the available range that is important, not the magnitude of the range.

A question may be raised as to the number of parameters that should be included in the calculation of the dissimilarities between samples. Sneath and Sokal (49) indicate that the greater the number of parameters included in the calculation, the more reliable the classification will be. To try to pick a subset of the data may bias the result. Certain elements may provide better discrimination between certain groups at the expense of poorer separation of others. The advantage of multielemental techniques like instrumental neutron activation or x-ray fluorescence is that they can provide the analysis for a large number of elements and it is best to use as many as possible.

The distance measures, cos θ or the correlation about the mean can be calculated using the program DSTCMP (53). It must be remembered that the distance parameters are dissimilarity measures while cos θ and the correlation about the mean both indicate increasing similarity with increasing value. They must be transformed before being used in a clustering routine based on dissimilarities. The matrix of dissimilarity measures is then passed to a program that will systematically group the individual points into clusters and then combine clusters until all the points are combined into a single large cluster.

However, several different criteria can be used to determine which points or clusters to combine. One of the most versatile clustering programs is AGCLUS (56), which provides seven different clustering criteria. These criteria can be divided into those that cluster points or groups by the distance between the groups and those that define the size of the cluster which results from the clustering process.

The "nearest neighbor" criterion is that the clusters are combined which have the minimum of the dissimilarities between points a in cluster A and points b is cluster B.

$$C_1(A,B) = \min d(a,b) \qquad (40)$$

where the minimum is over all points a in A and b in B. This minimum between cluster distance is supported by Jardine and Sibson (57) and is criticized by Lance and Williams (58).

Alternatively, the measure of the distance between clusters A and B can be the largest of the dissimilarities between the points of A and the points of B.

$$C_2(A,B) = \max d(a,b) \qquad (41)$$

where the maximum is over all points a in A and b in B. The clusters with the smallest values of C_2 are then combined. The same result is obtained by defining the size of cluster A as the maximum of the dissimilarities between pairs of points in A:

$$S_2(A) = \max d(a,a') \qquad (42)$$

where the maximum is over all of the points a and a' in the combined cluster A. This criterion is quite popular in the behavioral sciences. It is referred to by Johnson (59) as the cluster diameter. It can also be called the maximum between cluster distance or maximum within cluster distance.

Instead of using either the minimum or maximum of the distances between clusters, the average distance can be employed as the criterion.

$$S_3(A,B) = \frac{1}{n_1 n_2} \sum d(a,b) \qquad (43)$$

where n_1 is the number of points in A, n_2 is the number of points in B, and the summation is over all points a in A and b in B. This measure would not be as sensitive to a badly determined single parameter that might skew the results.

The other criteria depend on the measures of the size of the resulting cluster following combination. The object is to combine the clusters that result in the smallest value of the clustering criterion. The most straightforward of these size parameters is the mean of the distances between all pairs of distinct points within the cluster.

$$C_4(A) = \frac{2}{n_A(n_A - 1)} \sum_{a \neq a} d(a,a') \qquad (44)$$

where n_A is the number of points in A and the summation is overall pairs of points in A such that $a \neq a'$. If, for example, the input matrix were correlation coefficients, then $C_4(A)$ would be the mean inter-item correlation.

This criterion can be modified by allowing the summations of all pairs of points in A so that the diagonal elements $d(a,a')$ are included in the criterion.

$$C_5(A) = \frac{1}{n_A{}^2} \sum d(a,a') \qquad (45)$$

where now the summation is over all points in A. The distance measures between a point and itself will be zero while the other measures, with the exception of the MCD, will all be greater than zero. Thus, in most cases, the effect of using this criterion will be to decrease the apparent cluster size, and this reduction will be relatively larger for smaller clusters. This criterion biases the results toward smaller clusters. However, this criterion has an interesting feature when used in conjunction with the SED. For this case the cluster size is the mean squared distance of the points from the centroid of the cluster and the hierarchical clustering method tries to minimize this cluster spread. This cluster size could also be considered as the within-cluster variance, which is being minimized as clusters are combined.

If $C_5(A)$ is multiplied by n_A, the next criterion is obtained:

$$C_6(A) = n_A C_5(A) = \frac{1}{n_A} d(a,a') \qquad (46)$$

where the summation is the same as for C_5. This criterion will bias the clustering even more heavily toward clusters with small numbers of points. The interesting use of C_6 is again with the SED dissimilarity measure. The cluster size becomes the within-cluster sum of the squares.

Finally, a criterion can be established that calculates a distance between clusters A and B based on the increase in the total size of the combined cluster C over the sizes of A and B.

$$C_7(A,B) = C_6(C) - C_6(A) - C_6(B) \tag{47}$$

This criterion does not have the bias toward small numbers of points that C_6 does. Again using the SED measure, the criterion is then the increase in within-cluster sum of the squares that comes from merging A and B. This increase would be minimized at each stage in the sequential clustering process.

There are then several different dissimilarity or similarity measures and several ways of combining the points into clusters based on criteria regarding the measures. How does one choose the "best" measure and clustering criterion? It is impossible to provide a general answer applicable to all problems. In some cases it may be possible to decide *a priori*, but, in most cases, it is necessary to experiment with the possibilities until some experience is gained for various kinds of problems. In previous work on environmental trace elemental data, Hopke (12, 13) has found that the SED and clustering criterions C_5 to be the most useful in providing interpretable clusters of the data points.

To illustrate the use of cluster analysis, the results of several trace elemental investigations will be presented. Hopke and co-workers (60) reported the chemical and physical parameters describing the nature of grab samples of sediment taken from a small lake in southwestern New York State. In an effort to interpret this data set of 15 elements determined for each sediment sample, a cluster analysis using SED and C_5 was made (12) and the resulting pattern, called a dendogram, is given in Figure 1. The numbers used to label the points are the site numbers given in Reference 60. It can be observed that the data fall into four main groups: A, B, C, and D. Group A is clearly different from the others and groups B and C are more similar than D. An examination of the location of the sites in the lake where the samples were taken can be made using Figure 2. All the sites in group A are in the center of the lake while clusters B, C, and D are near shore sites. To assist in interpreting the differences among

Chautauqua Lake Sediment Analysis

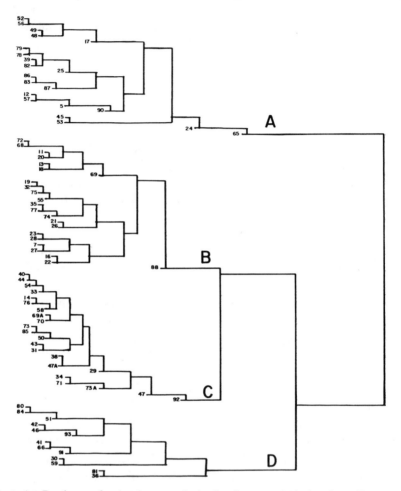

Figure 1. Dendogram for the cluster analysis of sediment analysis data from Chautauqua Lake, New York. Labels on the left side represent the site numbers assigned to each location by Hopke et al. (1976).

these groups, the average values were calculated for each of the elemental concentrations and for the parameters describing the particle size distributions and are given in Table 17.

Sediments from group A sites have a high clay fraction, are rich in arsenic, bromine, cesium, antimony, manganese, iron, potassium and the

TABLE 17. Average Values for Clusters for Chautauqua Lake Sediment Data

Variable	Cluster A	B	C	D
Arsenic	31.5	19.5	9.5	5.2
Bromine	24.5	10.0	4.4	1.7
Cesium	12.3	5.7	2.8	1.4
Europium	0.97	0.72	0.68	0.18
Iron (%)	12.34	10.02	4.78	3.05
Gallium	22.1	19.5	12.0	7.32
Hafnium	14.4	13.0	18.3	18.4
Potassium (%)	1.67	1.21	0.92	0.93
Lathamium	35.7	27.6	17.5	9.4
Manganese	2193.	1429.	485.	288.
Sodium	4194.	5604.	7576.	7465.
Antimony	6.8	2.3	1.6	0.8
Scandium	66.1	30.5	23.0	19.0
Tantalum	2.3	1.6	1.3	0.9
Terbium	1.3	0.7	0.5	0.4
% Sand	1.81	6.31	40.01	80.73
% Silt	51.10	62.16	44.43	11.54
% Clay	36.36	20.94	9.08	3.78
% Organic carbon	10.73	10.59	6.48	3.95
Depth	8.58	4.00	2.97	2.28
Median grain size (mm)	0.730	1.99	8.07	22.30
Mean grain size (mm)	0.595	1.26	5.91	21.34
(Median grain size)^2	0.613	4.54	103.4	543.3
(Mean grain size)^2	0.381	1.82	54.2	506.5
Median grain size (0)	7.24	5.77	3.96	2.23
Mean grain size (0)	7.45	6.41	4.35	2.32
Sorting	2.23	2.18	2.00	1.58
Skewness	0.136	0.424	0.367	0.273
Kurtosis	0.703	0.937	0.650	1.954
N. Kurtosis	0.413	0.478	0.608	0.60
(Median grain size)$^{-1}$	1.747	0.592	0.184	0.049
(Mean grain size)$^{-1}$	1.860	0.908	0.234	0.054

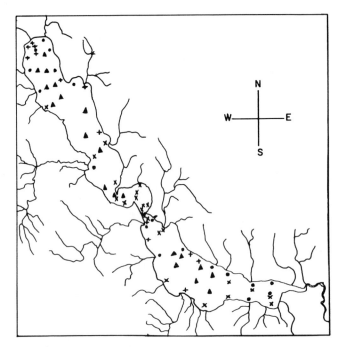

Figure 2. Map of Chautauqua Lake with site locations indicated to show to which cluster they belong. The symbols are defined as Δ for cluster A, x for cluster B, ○ for cluster C and + for cluster D.

rare earths. They are depleted in sodium. The sediments from group D are very sandy, have the highest sodium and the lowest organic carbon content. These are sandy beach areas which support very little plant growth. The B and C sites are of an intermediate nature between the sand of group D and the clay of group A. B sites tend to be in the deltas of the larger streams feeding the lake. The streams carry fine-grained material that is deposited in the deltas and then gradually spread by wave and current action to the C sites. The cluster analysis has provided clear consistent groupings of the sites and allowed for an initial understanding of the movement of material in this system.

Hopke and co-workers (61) have used cluster analysis to group their data on concentrations of 18 elements found in the urban aerosol of the Boston metropolitan area. In that case there was a considerable difference found between samples depending primarily on location of the sampling site. An analysis made on airborne particulate concentrations measured around the electric generating station in rural Chalk Point, Maryland, (34) showed the clustering primarily dependent on the day the samples were taken and very little effect of location. Thus the majority of the particulate mass is coming into the region with elemental concentrations relatively uniformly distributed. The origin of the particles is dependent on the prevailing wind direction, and the particles collected during one day have similar concentrations at all sites. The cluster analysis pattern can often provide a number of insights into what factors are important in causing the observed variations in a set of data by separating samples into groups having similar properties. By looking at the average values of the parameters used to describe the system, the differences between groups can generally be understood.

5. CONCLUSIONS

Two multivariate analysis techniques have been described and their usefulness in interpreting environmental data has been illustrated with several examples. These techniques probably will be more widely employed to provide useful insights into the vast arrays of data already taken in a number of monitoring programs. Data are really quite useless without interpretation ultimately leading to an understanding of the physical and chemical interactions that occur in environmental systems. These multivariate techniques can prove to be important tools in developing an in-depth understanding of environmental data and to make it easier to improve the quality of the human environment.

REFERENCES

1. D. W. Schindler, *Science* **192**, 509 (1976).
2. S. Wold and K. Andersson, *J. Chromatogr.* **80**, 43 (1973).
3. R. B. Selzer and D. G. Howery, *J. Chromatogr.* **115**, 139 (1975).
4. P. H. Weiner, E. R. Malinowski, and A. R. Levinstone, *J. Phys. Chem.* **74**, 4537 (1970).
5. P. H. Weiner and E. R. Malinowski, *J. Phys. Chem.* **75**, 1207 (1971).
6. P. H. Weiner and E. R. Malinowski, *J. Phys. Chem.* **75**, 3160 (1971).
7. R. W. Rozett and E. M. Petersen, *Anal. Chem.* **47**, 1301 (1975).

8. G. L. Ritter, S. R. Lowery, T. L. Isenhour, and C. L. Wilkins, *Anal. Chem.* **48**, 591 (1976).

9. E. R. Malinowski and M. McCue, *Anal. Chem.* **49**, 284 (1977).

10. J. S. Mattson, C. S. Mattson, M. J. Spencer, and S. A. Starks, *Anal. Chem.* **49**, 297 (1977).

11. I. H. Blifford, Jr. and G. O. Meeker, *Atmos. Environ.* **1**, 147 (1967).

12. P. K. Hopke, *J. Environ. Sci. Health* **A11**, 367 (1976).

13. P. K. Hopke, E. S. Gladney, G. E. Gordon, W. H. Zoller, and A. G. Jones, *Atmos. Environ.* **10**, 1015 (1976).

14. P. D. Gaarenstroom, S. P. Perone, and J. L. Moyers, *Environ. Sci. Technol.* **11**, 795 (1977).

15. D. F. Gatz, *J. Appl. Met.* **17**, 600 (1978).

16. H. Sievering, M. Dave, D. Dolske, and P. McCoy, *Atmos. Environ.* **14**, 39 (1980).

17. C. W. Lewis and E. S. Macias, *Atmos. Environ.* **14**, 185 (1980).

18. D. J. Alpert, and P. K. Hopke, *Atmos. Environ.* **14**, 1137 (1980).

19. D. J. Alpert, and P. K. Hopke, *Atmos. Environ.* **15**, 675 (1981).

20. N. Z. Herdam, *Atmos. Environ.* **15**, 1421 (1981).

21. P. P. Parekh and L. Husain, *Atmos. Environ.* **15**, 1717 (1981).

22. R. C. Henry and G. M. Hidy, *Atmos. Environ.* **13**, 1581 (1979).

23. W. M. Cox and J. Clark, *JAPCA* **31**, 762 (1981).

24. R. L. Tanner and B. P. Leaderer, *Atmos. Environ.* **15**, (1981).

25. D. L. Duewer, B. R. Kowalski, and J. L. Fasching, *Anal. Chem.* **48**, 2002 (1976).

26. J. Imbrie, Factor Analysis and Vector Analysis Programs for Analyzing Geological Data, Technical Report No. 6, ONR Task No. 389-135, Northwestern University, Evanston, Illinois (unpublished).

27. H. H. Harmon, *Modern Factor Analysis*, 3rd ed., University of Chicago Press, Chicago, 1976.

28. E. R. Malinowski and D. G. Howery, *Factor Analysis in Chemistry*, Wiley, New York, 1980.

29. R. B. Cattell, *Handbook of Multivariate Experimental Psychology*, Rand McNally, Chicago, 1966, pp. 174–243.

30. L. Guttman, *Psychometrika* **19**, 149 (1954).

31. H. F. Kaiser and S. Hunka, *Ed. Psych. Measurement* **33**, 99 (1973).

32. R. E. Blackith and R. A. Reyment, *Multivariate Morphometrics*, Academic, London, 1971, pp. 201–211.

33. H. F. Kaiser, *Ed. Psych. Measurement* **19**, 413 (1959).

34. P. K. Hopke, The Application of Factor Analysis to Urban Aerosol Source Resolution, in *Atmospheric Aerosol: Source/Air Quality Relationships*, E.

S. Macias and P. K. Hopke, Eds., Symposium Series No. 167, American Chemical Society, Washington, D.C., 1981.

35. P. K. Hopke, Comment on Trace Element Concentrations in Summer Aerosols at Rural Sites in new York State and Their Possible Sources, by P. P. Parekh and L. Hussain, *Atmos. Environ.* **16**, 1279 (1982).

36. B. A. Roscoe, P. K. Hopke, S. L. Dattner, and J. M. Jenks, *J. Air Pollut. Contr. Assoc.,* **32**, 637 (1982).

37. H. R. Bowman, F. Asaro, and I. Perlman, *J. Geol.* **81**, 312 (1973).

38. S. L. Dattner and J. M. Jenks, presented to the Seventy-fourth Meeting of the Air Pollution Control Association, Philadephia, Pennsylvania (1981).

39. C. K. Liu, B. A. Roscoe, K. G. Severin, and P. K. Hopke, *Am. Ind. Hyg. Assoc.* **43**, 314 (1982).

40. P. K. Hopke, R. E. Lamb, and D. F. S. Natusch, *Environ. Sci. Technol.* **14**, 164 (1980).

41. B. A. Roscoe and P. K. Hopke, *Comput. Chem.* **5**, 1 (1981).

42. B. A. Roscoe and P. K. Hopke, *Anal. Chim. Acta* **132**, 89 (1981).

43. B. A. Roscoe and P. K. Hopke, presented at the ANS-ACS Topical Meeting of Atomic and Nuclear Methods in Fossil Energy Research, Mayaguez, Puerto Rico (1980).

44. B. A. Roscoe and P. K. Hopke, presented at the Fourth International Conference on Nuclear Methods in Environmental and Energy Research, Columbia, Missouri (1980).

45. O. Exner, *Collect. Czech. Chem. Commun.* **31**, 3222 (1966).

46. F. Mosteller, *Rev. Int. Int. Statis.* **39**, 363 (1971).

47. F. Mosteller and J. Tuckey, *Data Analysis and Regression*, Addison-Wesley, Reading, MA, 1977.

48. A. A. Clifford, *Multivariate Error Analysis*, Applied Science Publishers, London, 1973.

49. P. H. A. Sneath and R. R. Sokal, *Numerical Taxonomy*, Freeman, San Francisco, 1973.

50. F. R. Hodson, *World Archeology* **1**, 90 (1969).

51. G. K. Ward, *Archaeometry* **16**, 41 (1974).

52. B. Ottaway, *Archaeometry* **16**, 221 (1974).

53. A. M. Bieber, Jr., D. W. Brooks, G. Harbottle, and E. V. Sayre, *Archaeometry* **18**, 69 (1976).

54. B. R. Kowalski, T. F. Schatzki, and F. H. Stross, *Anal. Chem.* **44**, 2177 (1972).

55. T. H. Charlton, D. C. Grove and P. K. Hopke, *Science* **201**, 807 (1978).

56. D. C. Oliver, Aggregative Hierarchical Clustering Program write-up, preliminary version, National Bureau of Economic Research, Cambridge, MA (1973).

57. N. Jardine and R. Sibson, *Mathematical Taxonomy*, Wiley, New York, 1971.

58. G. N. Lance and W. T. Williams, *Comput. J.* **9,** 373 (1967).

59. S. C. Johnson, *Psychometrika* **32,** 241 (1967).

60. P. K. Hopke, D. F. Ruppert, P. R. Clute, W. J. Metzger, and D. F. Crowley, *J. Radioanal. Chem.* **29,** 39 (1976).

61. P. K. Hopke, E. S. Gladney, W. H. Zoller, and G. E. Gordon, Am. Chem. Soc. Div. Environ. Chem. Prepr. **16**(2), 13 (1976).

INDEX